Nicholas Senn

Tuberculosis of the genito-urinary Organs

Male and Female

Nicholas Senn

Tuberculosis of the genito-urinary Organs
Male and Female

ISBN/EAN: 9783337140168

Printed in Europe, USA, Canada, Australia, Japan

Cover: Foto ©berggeist007 / pixelio.de

More available books at **www.hansebooks.com**

TUBERCULOSIS

OF THE

GENITO-URINARY ORGANS,

MALE AND FEMALE.

BY

N. SENN, M.D., Ph.D., LL.D.,

PROFESSOR OF PRACTICE OF SURGERY AND CLINICAL SURGERY, RUSH MEDICAL COLLEGE;
ATTENDING SURGEON TO PRESBYTERIAN HOSPITAL; SURGEON-IN-CHIEF,
ST. JOSEPH'S HOSPITAL, CHICAGO.

ILLUSTRATED.

PHILADELPHIA:

W. B. SAUNDERS,

925 WALNUT STREET.

1897.

Copyright, 1897, by
W. B. SAUNDERS.

PREFACE.

TUBERCULOSIS of the male and female genito-urinary organs is such a frequent, distressing, and fatal affection that a special treatise on this subject at the present time appears to fill a gap in medical literature. The different forms of genito-urinary tuberculosis as it occurs in both sexes come under the observation of the physician, surgeon, and gynecologist, and it is therefore a subject that must necessarily interest them equally. In the preparation of this book it has been the object of the author to place the available clinical material upon an etiologico-pathological basis. The bacteriology of tubercular affections of the genito-urinary tracts has also received due attention. The modern diagnostic resources employed in the differential diagnosis between tubercular and other inflammatory affections of the genital and urinary organs have been mentioned, and when deemed necessary are fully described. The medical and surgical therapeutics of the affections of which this book treats are at this time not in a satisfactory state, but the opinions and views of surgeons of large experience have been freely quoted, and in appropriate places the author has related the results of his own clinical observations.

N. SENN.

CHICAGO, August 1, 1897.

CONTENTS.

PART I.
                                                          PAGE

TUBERCULOSIS OF THE MALE GENITAL ORGANS . . . 17

    Tuberculosis of the Penis, 19. Tuberculosis of the Urethra, 28. Tuberculosis of the Spermatic Cord, 33. Tuberculosis of the Seminal Vesicles, 34. Tuberculosis of the Prostate, 40.

PART II.

TUBERCULOSIS OF THE TESTICLE AND EPIDIDYMIS   47

PART III.

TUBERCULOSIS OF THE FEMALE ORGANS OF GENERATION . 82

    Manner of Infection, 84. General Remarks on the Pathology of Tuberculosis of the Female Generative Organs, 93.

PART IV.

TUBERCULOSIS OF THE VULVA 102

PART V.

TUBERCULOSIS OF THE VAGINA . . 111

PART VI.

TUBERCULOSIS OF THE UTERUS 122

PART VII.

TUBERCULOSIS OF THE FALLOPIAN TUBES . . 149

PART VIII.

TUBERCULOSIS OF THE OVARY 173

PART IX.

TUBERCULOSIS OF THE BLADDER 185

PART X.

TUBERCULOSIS OF THE KIDNEY 222
   Miliary Tuberculosis, 233. Caseous Nephritis, 235. Tubercular Pyelonephritis, 240. Operations for Tuberculosis of the Kidney, 284. Nephrotomy, 288. Nephrectomy, 292. Subcapsular Nephrectomy, 307. Partial Nephrectomy, 308. Laparonephrectomy, 309.

TUBERCULOSIS
OF THE
GENITO-URINARY ORGANS,
MALE AND FEMALE.

PART I.

TUBERCULOSIS OF THE MALE GENITAL ORGANS.

The male genital organs are the seat of as yet imperfectly understood conditions that predispose them to tubercular infection. The literature on tuberculosis of these organs is scanty as compared with that of tubercular affections of other organs, such as the lungs, pleuræ, peritoneum, lymphatic glands, bones, joints, meninges, and skin. A careful search of what has been written on tuberculosis of the male genital organs will convince the searcher for truth and instruction that this subject has not received the attention its importance demands. The observations of many clinicians in this comparatively new field of surgical pathology present no uniform, and many times distinctly opposite, results. The deductions drawn in the post-mortem room by professional pathologists likewise lack uniformity, and the different results obtained have often been taken as a basis with which to fortify the opinions of individual surgeons. This department of surgical tuberculosis is in its primitive stage, and offers many inducements and opportunities for careful clinical observation and bacteriological and pathological research in the future. The great obstacle to a more perfect development of the surgical aspects of tuberculosis of the male organs of generation in the past, and to a certain extent at the present time, has been, and remains, accuracy in diagnosis. Since the mystic term "scrofula" has been almost completely eliminated from the present nomenclature of surgical affections it has been found that many of the chronic inflammatory diseases of the male organs of generation are of a tubercular nature. The important question as to whether the male genital organs are the seat of primary tuberculosis, or whether the disease extends to them by a progressive infective process from the upper portion of the urinary system, has not been definitely settled. Each side of the question can show its exponents whose views command respect. For my own part, I am firmly convinced that in a fair percentage of cases the male genital organs are the seat of primary tuberculosis, the tubercle bacilli finding in the blood-vessels of the complicated genital apparatus a condition favorable for their mural implantation, growth, and reproduction. These are the comparatively rare cases of hematogenous primary tuberculosis of the male genital organs, caused by the deposition, growth, and reproduction of tubercle bacilli floating in the general circulation, locating in some part of the genital apparatus, without a discoverable tubercular lesion in any of the other organs. Such cases occur, but are few as compared with instances of secondary tuberculosis complicating tubercular affections of other organs or occurring in the course of extension of a tubercular process by continuity of surface from the urinary organs. It is my purpose in these pages to call attention to some of the salient points in the etiology, pathology, and clinical aspects of tuberculosis of the male genital organs as set forth in current surgical literature, and I shall emphasize some of the topics that have attracted my attention by my own personal observations.

Tuberculosis of the Penis.—The frequency with which tubercular affections of the female organs of generation occur has recently received the well-merited attention of gynecologists. The observations that have been made in this direction have reminded the surgeon of the possibility of direct inoculation during coitus. That such an occurrence is beyond the range of imagination no one can deny. Years ago, Verneuil[1] expressed the opinion that primary genital tuberculosis which does not depend on scrofula is probably caused by direct infection during coitus—that is, by the wandering of tubercle bacilli through the external genital organs to a point of the apparatus in which conditions favorable for their localization and reproduction exist.

Poncet[2] reported to the French Congress for the Study of Tuberculosis an article on "Tuberculosis Having its Origin in the Penis." Three varieties are mentioned: 1. Balano-preputial tuberculosis; 2. Tuberculosis of the mucous membrane (this variety usually showing itself first in the deep urethra); and 3. A tuberculosis of the urethra that

---
[1] "Hypothèse sur l'Origine de Certaines Tuberculeuses Génitales dans les deux Sexes." *Gaz. hebd.*, 1883, Nos. 14, 15.
[2] *La Médecine Moderne.* Paris, July 29, 1890.

consists of fungous masses involving the peri-urethral tissues, thereby allowing the urine to infiltrate the penile structures. Kraske has observed on the dorsum of the glans penis of a patient aged forty-nine years a tubercular ulcer which was well demonstrated microscopically after amputation.

I have reason to believe that in many of the cases of destructive lesions of the penis treated by amputation of the organ, and in which no recurrence followed the operation, the disease was not carcinoma, as surmised, but tuberculosis. During the last two years I have seen two cases of extensive destruction of the penis from what I believed to be a tubercular process. One of the patients was a colored man aged about thirty-five years, single, with no history of syphilitic infection. The ulceration commenced upon the external surface of the prepuce several years ago, and was not attended in the beginning by enlargement of the inguinal glands. The ulceration and sloughing extended successively to the glans and the body of the penis, and finally resulted in almost complete destruction of the entire organ. Later the inguinal glands and the skin covering the scrotum became involved. The inguinal glands became caseous, and several well-marked tubercular abscesses developed. When I saw the patient he was confined to his bed and the discharge from the extensive ulcerated surfaces had resulted in inoculation-tuberculosis which covered a considerable area of the gluteal region on both sides. Some of the ulcers would heal from time to time, after which the new scar-tissue would again break down and give place to an ulcerative process. No signs of syphilis could be detected upon any part of the surface of the body or in any of the internal organs. The patient had been subjected repeatedly to antisyphilitic treatment with various preparations of mercury and iodin without any improvement; in fact, such treatment appeared to aggravate the local conditions and still further impair the general health of the patient. Under antitubercular treatment, local and general, consisting in the use of antiseptics, balsam of Peru, and later iodoform, repeated curettage, and the internal administration of guaiacol and cod-liver oil, the disease was arrested, the ulcers healed rapidly, and the general health, which had been precarious for several years, was so much improved in the course of a few months that the patient was able to resume light work. I had no opportunity to make a microscopic examination of the tissues or to search for the bacillus in this case, but I have but little doubt, both from the clinical history of the case and the character of the local lesions, that the affection was one of primary tuberculosis of the skin which extended rapidly to all the tissues of the penis, later to the lymphatics, and finally to neighboring parts exposed to contamination from the profuse discharges from the ulcerated surfaces. The very fact that the lymphatic glands were converted in whole or in part into cheesy masses speaks for the diagnosis of tuberculosis and against that of syphilis. It is well known that in the negro tuberculosis often pursues an exceedingly rapid course. In a few weeks tubercular glands break down and suppurate —a condition often associated with quite extensive phlegmonous inflammation of the surrounding connective tissue, the consequence of a mixed infection with pyogenic organisms. It is therefore not surprising that in rare cases tuberculosis of the penis should result in extensive destruction or complete loss of the organ under the influence of a double infection in persons peculiarly susceptible to the ravages of this disease.

For a full report of this very interesting case I am indebted to Prof. Scales and Dr. Fondé.

John Mitchell, mulatto, age 31. Entered City Hospital, Mobile, December 22, 1893, with an ulcer on the balanopreputial fold and extending on the glans penis.

The ulcer was cauterized with nitric acid and an iodoform dressing applied. The patient improved slightly, and at the end of two months left the hospital. He returned, after an absence of six months, with the entire penis destroyed, and with a large ulcer on the nates which rapidly extended. He suffered great pain, showed serious pulmonary complications, was greatly emaciated, and was confined to bed for six months. During this time the actual cautery was applied, followed by dressings of a 5 per cent. solution of balsam of Peru in castor oil, and also a short course of guaiacol. After the first two cauterizations there was decided improvement in the ulcer, but the third and last application was followed by no benefit, the ulcer continuing to enlarge. He was, however, sufficiently improved to enable him to leave his bed for an hour at a time. The patient's condition remained about the same for four or five months, when he seemed steadily to grow worse. Dressings of iodoform were applied daily from this time until January 15, 1895, when Prof. N. Senn saw the case and delivered a clinical lecture in the hospital amphitheatre to the students of the Medical College of Alabama. The patient's condition was desperate at this time. The ulceration had extended by numerous ulcers closely adjoining and finally coalescing, until the entire perineal region, coccygeal fissure, and both nates were denuded, in some place the destruction extending deeply. The bladder was emptied through several perforations in the perineum, and single shallow ulcers appeared on the scrotum. The entire extent of the ulceration on the nates measured 10 inches vertically by 9 inches transversely, being greater on the left buttock.

Dr. Senn advised a course of guaiacol and tonics until the general condition of the patient should permit of a thorough and radical removal of all infected tissues by curettement.

Guaiacol was commenced in doses of gtt. v, well diluted in milk, three times a day, and the dosage was gradually increased until the patient was taking gtt. x four times daily. He was also given a course of syrup of iodide of iron. The patient improved sufficiently in three months to leave his bed and lounge around for most of the day; he slept and rested well, which he had not been able to do since his arrival at the hospital, unless under an opiate. He still suffered pain the greater part of the time.

Curettement May 31, Surgeon Wm. M. Mastin. The patient was anesthetized and the whole field of infection was carefully and deeply scraped out and the edges trimmed with scissors. Some of the pockets extended deeply into the nates and some burrowed far under the skin; one, at the gluteal crease, penetrated nearly to the femur in the adductor muscles. A dressing of iodoform was then applied. Rapid healing followed, and pain was almost entirely absent when the patient recovered from the immediate effects of the curettement.

Curettement July 10. The condition was very much better, and healing had taken place in considerable part of the field of the ulcer. Was again thoroughly curetted, the same marked relief and improvement following.

Curettement Aug. 7. An occasional fresh breaking down in the cicatricial tissue beneath the surface invariably yielded to the curette and healed rapidly.

Curettement Aug. 30, 1895. The same marked improvement. The main portion of the field showed healthy and pliable cicatrix.

Curettement Sept. 16, 1895. Treated as before, with same benefit.

Curettement Oct. 31, Surgeon T. S. Scales. There were several small and shallow ulcers in the coccygeal fissure and in the inguinal fold on each side the scrotum, in addition to the large and deep ulcer which remained in the gluteal crease; these ulcers were carefully scraped and iodoform thoroughly rubbed in. Improvement followed.

Curettement Jan. 20, 1896, Surgeon Jas. A. Abrahams. There were two remaining ulcers, one next the scrotum in the coccygeal fissure, and the other in the gluteal crease. These ulcers again received a careful and extensive removal of the involved tissue, iodoform being thoroughly rubbed in. This was followed by healthy granulation at the bottom of the ulcer and by rapid healing.

Curettement Feb. 20, 1896, Surgeon Jas. A. Abrahams. There was steady improvement; the only large ulcer remaining was the one in the gluteal crease, and it was filling up rapidly. This ulcer was scraped and treated as before, improvement following.

Since the second curettement the patient has been able to work around the hospital, and has acted as a nurse for some time. He is 6 feet tall and weighs 185 pounds. Pulmonary symptoms have disappeared, and he is robust, very strong, and healthy; he suffers no pain. The only remaining ulcer is the one at the gluteal crease; this ulcer is now shallow and about the size of a half dollar. This ulcer was scraped again, and finally healed permanently.

Very recently I have had an opportunity to examine a similar case in the service of Dr. Bouffleur at the Cook County Hospital. In this case the lymphatic glands in the groins became involved after a considerable portion of the penis had become destroyed by ulceration and sloughing. Syphilis was suspected, but the most energetic treatment made no impression on the progress of the disease. Local and general antitubercular treatment with complete excision of all the infected tissues effected a speedy and permanent cure. Many sections of the diseased tissue were examined for bacilli with negative results, but the existence of isolated multinuclear giant cells furnished an additional proof of the tubercular nature of the primary disease and the secondary glandular complications.

For a full report of this case I am indebted to Dr. Bouffleur.

Emil Schmidt, single, white, florist, a German, 21 years old, admitted to Cook County Hospital Sept. 18, 1895.

Family History.—Father, mother, one brother, and three sisters living and well. One brother died of consumption at 26, after an illness of one year. No other history of consumption in the family.

Previous History.—Has never been sick except with a "carbuncle" on the back, that healed rapidly, and with gonorrhea several years since.

No history of primary sclerosis, of ulceration, or of secondary or tertiary syphilitic lesions.

Present Illness.—Two years ago noticed a small pin-point, reddish, slightly painful depression in the sulcus just back of the corona glandis on the dorsum of the penis. The patient declares that he had not been exposed to venereal infection for six weeks previous to the first appearance of the lesion. A physician gave the patient a solution of "lunar caustic," which he applied to the parts himself three times daily. During the three weeks of this treatment the lesion ulcerated rapidly and extended laterally in the sulcus until it encircled about two-thirds of the circumference of the glans penis. The application of bismuth was then resorted to, but the parts became markedly swollen; caustics were then applied. Under this treatment there was established a process of repair to the extent of the adherence of the foreskin to the glans penis, but the ulcer has never entirely healed.

About four weeks after noticing the primary lesion the patient detected in the left groin a swelling the size of a hazelnut. It was tender and painful, and enlarged to the size of a walnut in about two weeks. After the application of poultices the swelling ruptured, but the patient says that only a dark bloody fluid was evacuated. During this treatment he had been in the county infirmary, and at the time of his leaving, in June, 1894, there was an ulcer about an inch long with thickened granulated borders. This ulcer remained about the same size for six months, when it began to enlarge. During the following six or seven months it spread so as to involve an irregular triangular area about six square inches in extent.

The patient's condition at time of entrance was that of general good health excepting a well-marked anemia. He was fairly well nourished; tongue was clean, appetite good, digestion good, bowels regular, urination normal. Examination of mouth, nose, throat, chest, and abdomen negative excepting an inguinal hernia which rarely descended. The patient wore a truss. No periosteal nodes and no inguinal adenopathy. There was no adenopathy detectable anywhere.

The foreskin was adherent to the glans penis; about two-thirds of the circumference of the glans and the prepuce was occupied by an irregular firm lesion with a definite outline beyond which there was no induration. In the left groin there was an irregular triangular ulcerative lesion with well-defined boundaries. The outline of the triangle was about $2\frac{1}{2}$ by 4 by 5 inches, with the base one inch above and parallel to the middle part of Poupart's ligament, and the apex in the fold between the scrotum and the thigh. There was no induration beyond the elevated edges of the lesion. There were several small areas of pale granulations which secreted slightly and from which emanated a peculiar odor. The paleness of the surface increased toward the center.

Microscopic examination of the tissues and implantation-experiments were made by the house surgeon, who elicited the foregoing history, with negative results. The parts were dressed antiseptically and the patient was given full doses of potassium iodide and tonics, but without any effect upon the course of the lesions.

On October 2 the patient was chloroformed and the ulcers carefully and thoroughly cleaned out with the curette. While this operation was followed by cicatrization in most parts of the wounds, it was only temporary, and in several places persisted as a chronic slightly discharging surface.

At the time of the last operation, January 16, the whole of the glans penis was the seat of an ulcer which had destroyed a large part of the glans. The foreskin was adherent about the base of the glans, and the border was greatly thickened, being as much as half an inch thick and very firm. The swelling was not edematous, but was firm and composed of cicatricial and hypertrophied tissue. There was no adenitis. The open parts of the bluish raised lesion in the groin were principally at the angles of the ulcer, the central parts being occupied by a grayish, poorly nourished cicatrix.

Curettement having failed, resort was had to the radical operation of complete extirpation *en masse* of both of the involved areas. The glans penis and the thickened foreskin were dissected from the urethra, which was slit and attached to the center of the line of sutures of the healthy foreskin, which was slid forward so as to cover completely the remainder of the glans penis. The large area in the groin was removed by an incision carried beyond the involved tissues on all sides and beneath all the diseased tissue, through the deeper part of the superficial fascia. An attempt was then made to cover the area by means of a transferred flap and lateral sutures of the secondary wound. Owing to a disturbance of the dressings, the wound became infected, resulting in necrosis of a part of the transposed flap. The wound of the penis healed kindly, and, while the large area left in the groin cicatrized slowly, it finally healed completely and permanently.

Microscopic Examination.—Unfortunately, all the specimens from the groin and most of those from the glans penis became misplaced in the process of hardening. Of the tissues left, a number of sections were made and stained for the bacillus of tuberculosis, with negative results. A few giant cells were found in some of the sections.

In both of these cases the base and borders of the ulcers were not much indurated, the surface was covered with pale, flabby, edematous granulations, the margins well undermined, and the overhanging skin was of a bluish tint. The spongy and cavernous portions of the penis appeared to yield alike to the tubercular destruction. In neither of these cases did the disease involve the urethral mucous membrane above the level of the ulcerated surface. The strongest argument in favor of the tubercular nature of the destructive process is the fact that vigorous antisyphilitic treatment not only failed to arrest the disease, but resulted in aggravation of the local conditions and impairment of the general health of both patients, while the antitubercular treatment yielded the most prompt and satisfactory results.

Tuberculosis of the Urethra.—Surgeons are familiar with the well-known clinical fact that foreign substances when introduced into the urethra are very prone to travel in the direction of the bladder unaided by any *vis a tergo*. It is reasonable to assume that micro-organisms lodged in the meatus are conveyed in a similar manner along the urethral tract, and unless they become arrested upon a soil propitious for their growth and development they produce no symptoms. There can be but little doubt that direct infection of the urinary tract with the bacillus of tuberculosis occasionally takes place in this manner. Primary tuberculosis of the urethra is exceedingly rare, and when it occurs it takes place in a part of the urethral mucous membrane prepared for the reception and growth of the bacillus by some antecedent injury or disease. Tuberculosis of the urethra must be mentioned especially as an affection that is prone to exist in cases of vesical and prostatic tuberculosis, in consequence of a direct extension of the infective process from the bladder or the prostate to the mucous membrane of the urethra. The disease may occur in the prostatic, the bulbous, or any other part of the urethra; it is found more frequently in young females suffering from bladder-tuberculosis than in men; it appears in the form of ulcers, and is often attended by incontinence of urine. What English has designated as "tubercular peri-urethritis" is in reality a tubercular perineal abscess, which may originate as well from a tubercular Cowper's gland as from a tubercular urethral ulcer. I have now under my care at the Presbyterian Hospital a boy fourteen years of age who about a year ago manifested the first symptoms of primary renal tuberculosis. In the course of a few months there developed on the left side a large tubercular paranephric abscess, which was incised, but which never healed. Soon after the boy was admitted to the hospital symptoms appeared indicating that the tubercular process had reached the bladder. A few weeks later there set in a urethritis characterized by a profuse purulent discharge. The meatus presented the same appearances as during the early stages of a gonorrheal urethritis. As soon as the disease reached the urethra incontinence of urine was manifested and continued until the acute symptoms subsided. Four guinea-pigs were inoculated by injecting a hypodermatic syringeful of pus either into the peritoneal cavity or into the loose connective tissue in the groin. All the animals died in the course of five to six weeks, and the post-mortem in each instance revealed diffuse miliary tuberculosis. In this case the tubercular inflammation extended from the kidney over the entire urinary tract in the course of a year. Long before the infection reached the bladder and urethra the tubercular nature of the primary renal affection was established by the detection of numerous tubercle bacilli in the sediment obtained from the centrifuge.

Tubercular urethritis occasionally develops in consequence of rupture into the urethra of a tubercular abscess of the prostate. Soloweitschick[1] describes a case in which the urethral affection had such an origin, the disease following tuberculosis of the testicles, vasa deferentia, seminal vesicles, and prostate. The latter contained a caseous cavity which had perforated into the urethra. The urethral mucous membrane was the seat of numerous chancre-like tubercular ulcers.

Tubercular urethritis gives rise most constantly to retention and incontinence of urine. Owing to the irritation caused by the urinary secretion, which will be voided the more frequently in proportion to the irritation of the neck of the bladder present, the inflamed mucous membrane will be kept in a constant state of disease, and the more so as in these situations the tubercular infiltration is not usually eliminated, but, on the contrary, steadily increases in quantity, and on this account not only excites catarrhal inflammation in the adjacent healthy mucous membrane, but also affords at the deepest part of the urethra an increasing impediment to the flow of urine, and the most appropriate local surgical treatment only suffices to check in some measure the retention of urine. Retention eventually leads to incontinence. Michaut[2] ob-

---

[1] *Archiv f. Dermatologie*, 1870.

[2] "Sur un cas d'ulcération tuberculeuse de l'urèthre consécutive à une tuberculose renale primitive." *Bull. de la Soc. Anat. de Paris*, 1887.

served a case of tubercular ulceration of the urethra in a man the subject of pulmonary tuberculosis. The disease appeared as a hard induration, five centimeters behind the meatus, which simulated closely a hard chancre. Later, tubercular granulations appeared around the meatus and upon the surface of the glans penis. The post-mortem revealed tuberculosis of the kidneys and a descending tubercular process which finally reached the urethra. The part of the urethra affected was indurated, and the fossa navicularis was the seat of deep ulcerations.

In the differential diagnosis of urethral chancres it is well to bear in mind tubercular lesions which may so closely resemble primary syphilitic infection. Analécot[1] records a rare case of secondary tubercular ulcer surrounding and involving the meatus, about the size of a ten-cent piece. The patient, a boy fourteen years old, had been circumcised eight days after birth. The appearance of the ulcer and the increase of induration excluded the idea of hard chancre. The smooth base, the absence of suppuration, and the regularity of the border excluded chancroid. The ulcer was not painful. The patient had for three years been suffering with bladder-trouble. Micturition was frequent and painful, and the urine was at times bloody. Although no bacilli could be detected in the ulcer, inoculation with débris taken from it nevertheless produced tuberculosis in guinea-pigs, and so demonstrated its tubercular nature. The ulcer had remained stationary for

---

[1] *Ann. des Maladies des Organes Génito-urinaires,* Nov. 1893.

nearly a year. Ahrens[1] succeeded in finding the reports of only four cases of tuberculosis of the urethra in women. It is more common in men, but its relative frequency is not estimated alike by different authors. Krzywicki believes that the urethra is affected in 1 per cent. of all forms of tuberculosis and in 17 per cent. of all cases of urogenital tuberculosis. In the majority of cases the urethra is affected secondarily both in the ascending and descending forms of urogenital tuberculosis, and with few exceptions the prostate gland is simultaneously affected. In exceptional cases urethral tuberculosis is met with as a primary affection, and in that event is nearly always mistaken for a primary syphilitic ulcer. In the primary form the infection takes place either through the general circulation or by inoculation. The latter mode of origin was demonstrated experimentally, to a certain extent at least, by Baumgarten by his experiments on rabbits. As a pathological curiosity must be mentioned tubercular stricture of the urethra. Such a case is described by Ahrens.[2] The patient was a boy sixteen years of age, who was at the same time the subject of tubercular coxitis. The stricture could be passed only with a filiform bougie. The patient died, six days after his admission into the hospital, in consequence of retentio urinæ and rupture of a diverticulum at the base of the bladder. The post-mortem showed a caseous exudate upon the surface in the posterior part of the urethra as far as the pars cavernosa.

---

[1] "Die Tuberculose der Harnröhre." *Beiträge zur klinischen Chirurgie*, B. viii. p. 312.

[2] *Op. cit.*

The bladder, testicles, seminal vesicles, ureters, and most of the internal organs were the seat of recent tubercular infection.

In primary tuberculosis of the urethra, when the disease is accessible, the most energetic local treatment should be resorted to with a view to eliminate the tubercular material; while its occurrence as a secondary affection to tubercular disease of other portions of the urogenital organs calls for palliatives and improvement of the general health of the patient by appropriate general treatment.

Tuberculosis of the Spermatic Cord.—The spermatic cord, or vas deferens, is the connecting channel between the testicle, the essential organ of generation in the male, and the seminal vesicle. It is never the seat of primary tuberculosis. In tuberculosis of the testicle the disease usually manifests an intrinsic tendency to advance in an upward direction, implicating the cord, and, if life is prolonged for a sufficient length of time, eventually reaching the seminal vesicles. The cord becomes enlarged, indurated, and usually nodular, so that when it is passed between two fingers it presents somewhat the outlines of a rosary. In some cases the cord enlarges to the size of the little finger. The swelling is either cylindrical, nodulated, or spindle-shaped. The mucous membrane is most thickened, then the muscular coat, and least the adventitia. Perforation of the wall leads to tubercular paraspermatitis. The tubercular process usually results in obliteration of the lumen of the cord. Obstruction takes place either by the accumulation of tubercular material or by cicatrization. The latter takes place after destruction of the mucous lining of the canal followed by the formation of new connective tissue. (Figs. 1 and 2.) Pain is usually absent, and tenderness on pressure is slight. In cases of primary tuberculosis of the seminal vesicles the infective process frequently descends along the cord to the epididymis.

The surgical interest in tubercular spermatitis centers in the operation for the removal of a tubercular testicle. In all cases in which the cord is affected the inguinal canal should be laid open freely as far as the internal inguinal ring, and by gentle traction and the use of dull instruments as much as possible of the cord should be made accessible and removed. It has been shown that with proper care the cord can be liberated in this manner and excised to a point very near the seminal vesicle.

Tuberculosis of the Seminal Vesicles.—The seminal vesicles are occasionally the seat of primary tuberculosis, but in the majority of cases the disease is associated with similar affections of other parts of the genital organs, most frequently the testicle and the prostate gland.

Guyon believes, with Lancereaux, that the tubercular process begins very frequently in the vesiculæ seminales. Of 26 autopsies made with reference to showing the frequency with which the seminal vesicles are primarily affected, Guyon found this to be the case in 2 cases; in 10 cases the seminal vesicles were involved, but the prostata was simultaneously affected; in 1 case the prostate alone was involved.

Of 36 cases of disease of the seminal vesicles

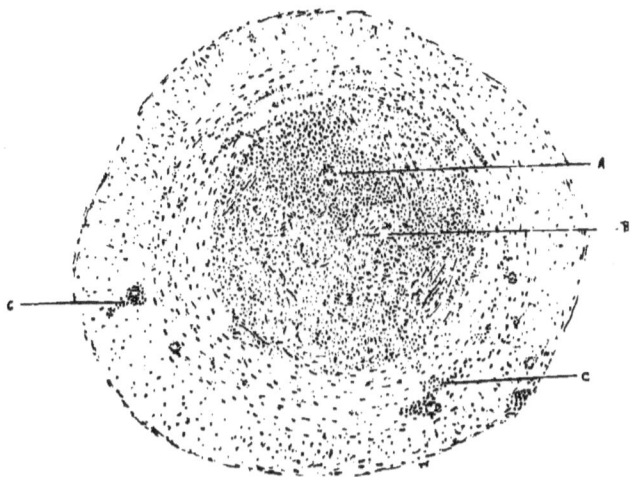

FIG. 1.—Transverse section of tubercular vas deferens, showing obliteration of the lumen by destruction of the mucous membrane and the accumulation of tubercular material; × 100: *a*, giant cell; *b*, obliterated lumen; *c*, infiltration and thickening of muscular coat.

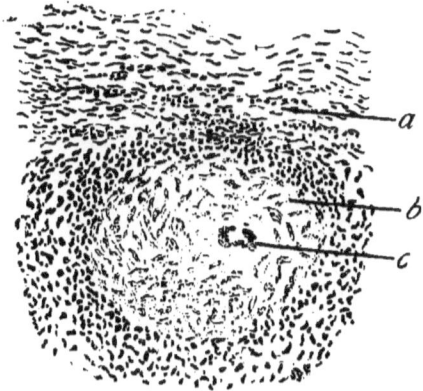

FIG. 2.—Histology of tuberculosis of the vas deferens; × 250: *a*, infiltration of muscular coat; *b*, epithelioid cells; *c*, giant cell in the center of a tubercle-nodule near the ulcerated mucous membrane and obliterated lumen of the cord.

collected by Dreyer,[1] in 18 the affection was of a tubercular nature. Of these cases, one-half occurred in persons over forty years of age. In 3 cases the vesicles were primarily affected; in 12 cases the disease presented itself at a stage in which the organs were hard and nodular without any softening; while in 6 cases the affection had passed into the second stage, characterized by caseation and liquefaction of the caseous material—that is, the formation of tubercular abscesses. In 3 cases the disease had extended beyond the capsule of the glands and had involved the pelvic connective tissue. In 1 case the disease was complicated by tubercular peritonitis. Pulmonary tuberculosis was absent only in 2 cases. As a rule, different portions of the urogenital tract and distant organs were found implicated.

Gueillot, in 59 cases of tuberculosis of the seminal vesicles that he collected from different sources, found involvement of the lungs 49 times, of the prostate 36 times, of the testicle and epididymis 34 times, and of the bladder and urethra 29 times.

The seminal vesicles can be palpated most satisfactorily by placing the patient in the knee-elbow position. In tuberculosis of one testicle the seminal vesicle on the corresponding side is frequently found affected, and in tuberculosis of both testicles the subsequent affection of the seminal vesicle is often bilateral. The existence of hard nodules in different parts of the organ which are not very tender on pressure is very suggestive of the tubercular nature of the disease. The close proximity of the seminal

---

[1] " Beiträge zur Pathologie der Samenbläschen." Inaugural Dissertation, Goettingen, 1891.

vesicles to the peritoneum in case these organs are tubercular must occasionally lead to peritoneal tuberculosis.

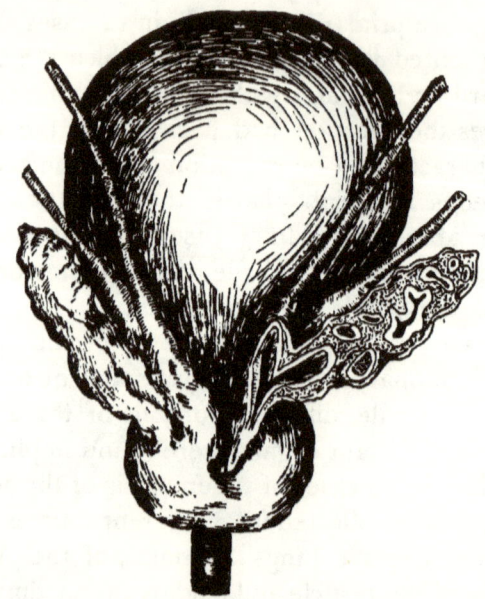

FIG. 3.—Tuberculosis of the seminal vesicles and the lowest portion of the vas deferens.

Gueillot believes that in rare cases tuberculosis of the seminal vesicles ends in recovery by transformation of the organ into a fibrous mass or by calcification. Broca found calcified tubercles in atrophic seminal vesicles. In such cases the adventitia is thickened, the muscular coat is the seat of fibrous degeneration, and the vesicles are sclerosed.

A number of surgeons have made bold attempts to eradicate one or both seminal vesicles by operative interference. Ullmann reports from Albert's clinic a case of extirpation of the tubercular vesiculæ seminales in a patient seventeen years of age who had been castrated for tubercular orchitis. Zuckerkandl's semilunar incision between the scrotum and the anus was made, the space between bladder and rectum exposed, and the posterior wall of the former made prominent by the use of a steel sound. The vesiculæ seminales and the vasa deferentia were now freely exposed and could be readily dissected out, as well as the upper left angle of the prostate, which contained a small abscess. Only the left, apparently healthy, vas deferens was left. The hemorrhage during the operation was free, and this, as well as the secondary hemorrhage which occurred in the evening of the same day, had to be arrested by a resort to the iodoform-gauze tampon. Healing of the wound took place quickly, with the exception of a small urinary fistula. The fistula was supposed to have been caused by division of the ejaculatory duct in the substance of the prostate. The patient left the hospital with a small urinary fistula, and claimed to have experienced erections at different times.

Ullmann regards primary tuberculosis of the seminal vesicles and unilateral secondary infection in the course of testicular tuberculosis as legitimate indications for a radical operation.

In two cases of secondary tubercular vesiculitis Roux[1] followed castration for tuberculosis of the testicle by excision of one of the seminal vesicles which had become involved by the tubercular process. After the removal of the diseased testicles

[1] "Extirpation de la vésicale séminale." *Congrès français de Chirurgie*, 1891.

and suturing of the wounds he brought the patient into the lithotomy position, lying on the affected side, and made an incision 2–3 centimeters from the median line as far as the ischium, exposing the rectum and penetrating finally as far as the affected vesicle. By pressure with the finger in the rectum from above downward the edge of the vesicle was made to appear in the upper portion of the wound. By the use of a traction ligature the vesicle was drawn further down, separated, and with the stump of the vas deferens was divided close to the neck of the bladder. The patient recovered from the immediate effects of the operation, but the remote results of the procedure are unknown.

Weir[1] reports a case in which he removed both seminal vesicles through Zuckerkandl's transverse perineal incision. Schede[2] removed successfully a tubercular seminal vesicle and cord. He prefers the method of Rydygier (which consists of a lateral incision extending along the border of the sacrum) to that of Dittel (which is the same as that which Schede employs for extirpation of the prostate— namely, Zuckerkandl's curved perineal incision).

Perhaps the most complete operation for tubercular vesiculitis has been performed by Fenger.[3]

The patient was 22 years of age, with a good family history. One year before his admission into the German Hospital, Chicago, he contracted a gonorrhea which persisted for six weeks. During the latter part of this disease he complained of rheumatic pains beginning in the left foot and passing thence to the hip

---

[1] *Medical Record*, Aug. 11, 1894.
[2] *Deutsche med. Wochenschrift*, Feb. 15, 1894.
[3] Personal communication.

on the same side, and later to the left knee and right ankle. At first the joints were not swollen, but about a week later the knee-joint began to swell and he was confined to his bed, where he remained for two months. At the end of this time he recovered almost completely from the first affection. In November, 1894, he noticed that the left testicle was swollen; the swelling increased slowly until in about four weeks it was as large as an adult's fist. When the patient came under Fenger's care the left testicle was firmer and slightly larger than the right. The epididymis was hard and nodular and slightly tender to pressure. Digital examination of the rectum showed no enlargement of the prostate, but the left seminal vesicle was felt as a hard mass, not tender to pressure. Reducible right inguinal hernia existed. The general health was not impaired. Operation April 25, 1894, in two steps: first, removal of testicle and accessible part of cord; second, extirpation of the seminal vesicle on the same side through Roux's incision. In removing the cord, after previous isolation of the testicle, the incision was carried as far as the internal inguinal ring, when the vessels were tied and the cord teased out of its canal as far as possible before dividing it. The wound was packed with iodoform gauze and sutures were introduced, but were not tied until after the completion of the second step of the operation. During this part of the operation the patient was turned on his right side, the knees being drawn up, and the rectum lightly packed with iodoform gauze. Roux's incision, four inches in length, was then made on the right side of the rectum. A sound was introduced into the bladder to serve as a guide in making the deep dissection. The seminal vesicle was located by the finger in the rectum as an olive-shaped body. Access to the vesicle was difficult, owing to the small size and great depth of the wound. Bleeding was not as profuse as was expected, and was easily controlled by the use of large hemostatic forceps. The seminal vesicle was found close to the bladder, and was drawn downward with Museux forceps. During the dissection the prostate and the levator ani muscle were seen and recognized. The vesicle was not encapsulated as was anticipated, and was accidentally opened and the contents of the tubercular abscess escaped. By careful dissection with Kocher's director and scissors the vesicle, with about $1\frac{1}{2}$ inches of the vas deferens, was removed without injuring the bladder or opening the peritoneal cavity. The prostate gland was found slightly enlarged. In the left lobe a whitish spot was seen; this was incised, and proved to be a small tubercular abscess. The abscess-cavity, with a portion of the lobe of the gland, was excised. The wound was sutured and was drained at each angle with tubular and gauze drains. The first wound was then closed and drained in a similar manner. Duration of operation, two and a quarter hours. The wounds healed kindly, and the patient remains in perfect health at the present time, more than two years after the operation.

This case appears to show more conclusively than any other on record that a timely operation for secondary tuberculosis of the seminal vesicle following an ascending tuberculosis of the testicle may succeed in preventing the extension of the disease to the bladder and other portions of the urinary tract, and may even result in a complete and permanent cure.

If the surgeon intends to remove both seminal vesicles, there can be but little doubt that Zuckerkandl's incision is the safest and renders the diseased organs more accessible than any other. The operation is greatly facilitated by placing the patient in the ventral position with the pelvis elevated.

Tuberculosis of the Prostate.—Sir Henry Thompson[1] is of the opinion that the prostate is never the seat of primary tuberculosis. He says: "It would appear that at no period of the disease is the prostate affected alone, some other part of the genito-urinary tract being the preliminary seat of the affection. In most cases the deposit appears

[1] *The Diseases of the Prostate: their Pathology and Treatment.* London, 1861, p. 283.

to take place first in the kidney, or at all events to be present there in an early stage. The organ next in order of liability to the disease, among the genito-urinary group, is the testicle. Thus, in 18 cases collected by myself, in which the results of post-mortem inspections have been recorded, tuberculosis of the kidney is reported in 13, and of the testicle in 7. The state of the lungs has, I suspect, not always been recorded, but in 10 of these cases they are stated to have been diseased."

Marwedel[1] has written a valuable monograph on tuberculosis of the prostate, in which he describes 4 cases that occurred in Czerny's clinic, in 2 of which cases the disease appeared as a primary tuberculosis of this organ; both cases were successfully treated by laying open and curetting the fistulous tract. In the other 2 cases temporary benefit resulted from incising the periprostatic abscess in front of the rectum, and through the incision the whole sequestrated gland could be removed. In these cases the destructive process had extended to the urethra, a considerable portion of which was involved. Another fatal case was not subjected to operative treatment, as it was complicated by pulmonary and testicular tuberculosis. These as well as other cases of a similar nature that have been recorded prove that the prostate may become the seat of primary tuberculosis.

Tubercular disease of the prostate is, as a rule, met with in young adults. Out of 26 cases col-

[1] "Aus der Heidelberger chirurgischen Klinik des Prof. Czerny, Ueber Prostata tuberculose." *Klinische Beiträge*, B. ix. p. 537.

lected by Socin, 13 were less than thirty years of age. The affection may, however, occur in men advanced in years. In one of Socin's cases the patient was seventy-two years old; he was suffering from the consequences of an enlarged prostate, and several months later the symptoms revealed extension of the tubercular process from the prostate to the seminal vesicles, bladder, and ureters, resulting in death after a long period of intense suffering.

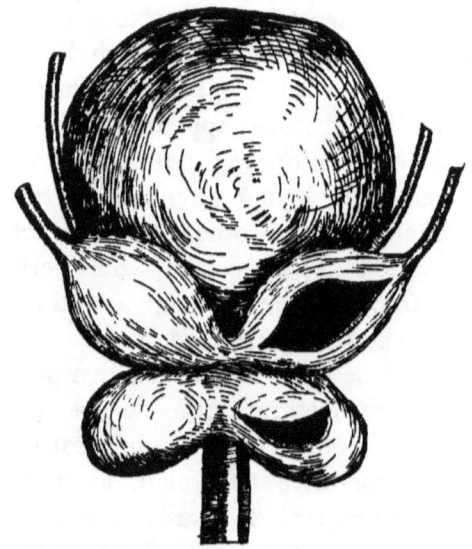

FIG. 4.—Tuberculosis of the prostate gland and the seminal vesicles.

The primary nodules are situated first in the vicinity of the tubules (Rindfleisch), and not, as was formerly supposed, in their interior. By confluence and caseation of the miliary gray nodules large masses and cavities are formed, which may be found in one or both lateral lobes, while the middle lobe is more rarely affected. The softening of the caseous material leads to the formation of tubercular abscesses, which may rupture into the urethra. In a case observed by Adams the abscess ruptured at the same time externally in the perineum. Socin observed two cases which terminated in rupture into the bladder, and in one case into the peritoneal cavity. The abscess may also discharge itself into the rectum.

Much less frequent than softening is calcification of the tubercular mass with arrest of the disease. Englisch[1] has made very valuable clinical observations concerning tubercular affections in the vicinity of the prostate gland and bladder independently of tuberculosis of these organs. The abscesses which form pursue a chronic course, and rupture finally either into the bladder, the rectum, or near the anus. He advises early incision only in cases where the tubercular infiltration involves the periprostatic plexus, when without prompt interference the danger is great from thrombosis of the hypogastric vein and pyemia. If the disease is extensive all around the bladder, incision through the abdominal wall is often the only means of preventing peritonitis.

It is difficult to ascertain whether or not tuberculosis of the prostate appears as a primary affection. Besides the cases cited above, Béraud and Pitha have observed cases in which this appears to have been the case. Post-mortem examination, however, usually reveals additional tubercular affec-

---

[1] "Ueber tuberculöse Infiltration des Zellgewebes in der Umgebung der Vorsteherdrüse und Blase." *Wiener Klinik*, 1896, Heft 1.

tions in adjacent or distant organs. In cases of tubercular abscess of the prostate it is difficult to ascertain the exact chronological order in the presence of additional tubercular lesions. Socin saw two cases of what appeared to be primary tuberculosis of the prostate. The patients were respectively twenty-eight and thirty years of age, and both of them had contracted gonorrhea before symptoms of prostatitis appeared. Examination revealed well-marked chronic prostatitis without additional tubercular complications. One of the patients died of an acute renal affection of only ten days' duration. The post-mortem showed disseminated cheesy foci in both lateral lobes of the prostate, ulcerative nephritis, and miliary tuberculosis of the peritoneum, the right pleura, and the liver. In the other case, months after the appearance of symptoms indicative of prostatitis and vesical catarrh, hectic fever set in, with symptoms pointing to renal disease, which rapidly proved fatal. The prostate was found completely destroyed, and its place was occupied by a tubercular abscess which communicated with the urethra. The vesical mucous membrane was the seat of numerous small ulcers, and at a point corresponding with the orifice of the left ureter was a deep ulcer; the left kidney was greatly enlarged; in the pelvis were large ulcers, in the substance of the kidney many small abscesses containing cheesy material. In two other cases the tubercular prostatitis occurred in phthisical patients. Socin is of the belief that in all these cases the prostatic disease was primary.

Of 26 cases of tuberculosis of the prostate collected by Socin, of which cases 6 came under his own observation, the post-mortem showed in 24 tubercular disease in other parts of the genito-urinary organs, and only in 2 cases tuberculosis of distant organs—the lungs and the bones. Most frequently the bladder and the kidneys were found implicated, less frequently one or both testicles. In one case in which the epididymis was unquestionably the primary seat of the tubercular process the lobe of the prostate on the corresponding side was similarly affected.

Symptoms.—There are no symptoms which are, strictly speaking, proper to this affection of the prostate. Undue frequency and pain in passing urine, occasionally blood in the urine, and at times the signs of cystitis, are commonly experienced. Wasting and extreme debility slowly show themselves. The symptoms present many things in common with other forms of chronic prostatitis. Only in cases in which the disease is complicated by tuberculosis of other organs are there observed hectic fever, rapid loss of strength, and marked emaciation. Adams has called attention to the similarity of the symptoms with those produced by stone in the bladder. Catheterization is always very painful, and should not be unnecessarily resorted to. In one case Socin observed, soon after catheterization, septic cystitis. Hematuria is often present, but is of no particular diagnostic value. Incontinence of urine, which often appears during the advanced stage, indicates extension of the disease to the sphincter vesicæ muscles. Marwedel found tubercle bacilli in the urine in all the four cases reported from Czerny's clinic; in one case they were also detected in the urethral discharge.

Treatment.—Czerny obtained good results in two cases which had terminated in abscess by laying open the fistulous tract and vigorously using the sharp spoon. In cases in which the disease has not advanced to abscess-formation he advises Zuckerkandl's perineal incision for exposing and removing the caseous foci.

Sir Henry Thompson advises conservative treatment, avoiding instrumentation, which in such cases provokes irritation and aggravates the disease without conferring upon the patient any benefit whatever.

Horteloup[1] recommends, in the treatment of isolated tuberculosis in any part of the genital tract, with a view to prevent further extension of the disease, the injection of a few drops of Lannelongue's solution of chloride of zinc. In two cases of tuberculosis of the prostate he made the injections through a *boutonnière* incision, but the communication was made too soon after the treatment was suspended to judge of its curative effects.

During the early stage of primary prostatic tuberculosis parenchymatous injections of iodoformglycerin emulsion would appear to be indicated, and might possibly prove a valuable addition to the treatment of this obstinate and unpromising affection. The diseased organ should be reached by a puncture through the perineum, and not through the rectum.

---

[1] "De la tuberculose génitale." *Gaz. Méd. de Paris*, No. 25, 1892.

PART II.

TUBERCULOSIS OF THE TESTICLE AND EPIDIDYMIS.

EXCEPT in cases of acute diffuse miliary tuberculosis, the essential organ of generation in man is seldom the seat of primary tuberculosis. On the other hand, primary localization of the tubercle bacillus not infrequently takes place in the epididymis, whence the infection frequently extends to the testicle by continuity of the tubercular process. There must be vascular conditions or other local predisposing causes which are concerned in determining hematogenous infection of a tubercular nature in the epididymis, which are absent or present to a less degree in the testicle. The epididymis is more frequently than the testicle the seat of acute infective processes which prepare the soil for the bacillus of tuberculosis; this fact may to a certain extent explain the greater frequency with which primary tuberculosis occurs in the former than in the latter. It must also be remembered that in descending tuberculosis from the upper portion of the urinary tract the epididymis is exposed first to infection, and that the patients often succumb to the primary disease and its complications before a sufficient time has elapsed for the testicle to become involved.

There still prevails the greatest diversity of opinion among pathologists and surgeons in regard to the epididymis being most frequently affected in cases of urogenital tuberculosis. Councilman[1] believes that in such cases the most common seat of the primary disease is the epididymis. He says: "It may be confined to this, or the testicle may be affected by continuity. The epididymis is converted into a more or less firm, caseous mass. From this the disease extends along the vas deferens, which becomes enlarged, and on section the interior is found to be lined with a whitish caseous tissue. In both the vas deferens and the epididymis the seat of the disease is primarily in the epithelium, and later takes the form of a tubercular inflammation. The seminal vesicles on the same side become affected in most cases, or they may be passed by and the disease appear in the prostate or the bladder. Up to this point it is easy to see how the infection has taken place. The extension has been in the direction of the secretion, and the bacilli could be carried along with the secretion. From the bladder the extension is in a direction opposite to the flow of the secretion; with or without any involvement of the ureter, infection of the pelvis of the kidney and of the adjoining kidney-tissue takes place. It is probable that the bacilli find suitable conditions for growth in the ureter, and grow along the walls, just as on the surface of a solid medium, until the pelvis

---

[1] *System of Surgery*, Dennis, vol. i. p. 246.

of the kidney is reached. There is no other way for infection to take place from the bladder to the kidney than along the ureter. There is no lymphatic or vascular connection. The proof that this is the usual route of infection in genito-urinary tuberculosis is shown by the certainty with which the disease can be traced step by step, and the extreme rarity of the disease in females as compared with males. In some cases the disease appears to be primary in the kidney, and the infection may take place in an opposite direction." My clinical experience corresponds with the views of Councilman, that in more than 50 per cent. of urogenital tuberculosis the disease has its primary starting-point in the epididymis. We shall see, in considering the etiology of this disease, that there are many authorities who take the opposite ground, and affirm that urogenital tuberculosis most frequently has its origin in the upper portion of the urinary tract.

Etiology.—Aievòli[1] made experiments on guinea-pigs by injecting into the testicle and epididymis tubercular material and pure cultures of the tubercle bacillus. Only in one case did he find tubercle bacilli in the lumen of the seminal ducts, but in all instances there was produced an intercanalicular proliferation, so that the walls of the canals were perforated and large masses with caseous centers were formed in which pseudo giant cells could be demonstrated. At some places an attempt at cure by sclerosis could be found at the same time, also tubercle-production in the vicinity of blood-vessels

[1] *Eriberto sur la tuberculosi di testiculo ed epididimo,* Morgagni, 1891, pp. 657, 728.

and the interstitial tissue. He believes that such an intercanalicular tuberculosis is possible without direct inoculation, as the bacilli may reach the interstitial tissue through the lymph-stream. The results of these experimental researches are closely allied with the observations of pathologists concerning the gross pathological anatomy of testicular tuberculosis.

The disease in the epididymis is caused frequently by a descending infection from the prostate and seminal vesicles, but it may originate in the epididymis primarily, as tubercle bacilli have been found on several occasions in the seminal ducts in healthy testicles in phthisical subjects. The process begins in most cases in the epididymis, in the form of conglomerate tubercles, which are conspicuous for the number and size of the giant cells. The tubercle elements are derived mostly from the interstitial connective tissue, but part of the product may be furnished by the epithelial cells and other tissues of the seminal ducts. Infection may extend along the urogenital canal from the kidney to the testicle; as a rule, however, tuberculosis of the testicle begins in the epididymis. Sometimes testicle and epididymis are affected simultaneously in cases of general miliary tuberculosis. "The fact that the spermatic artery divides when it reaches the epididymis may account for the localization of the disease in the latter organ; the slowing of the blood-current always favors bacterial growth. Infection may also occur through the vas deferens. The conditions for localization of the microbes after entrance into the urethra on their way to the vas deferens are not as favorable as in the latter organ" (White).

The predisposing causes are inherited soil, antecedent or coexisting disease of the testicle, and trauma. The disease begins more frequently in the globus major than at the opposite end of the organ. Later the testicle and its envelopes are invaded by direct extension of the infective process from the epididymis to those organs. Age appears to have a positive determining influence in the production of tuberculosis of the epididymis and testicle. Salleron ascertained the age in 47 cases of tuberculosis of the testicle with the following result: 20-30 years, 36; 30-40 years, 6; 40-50 years, 4; 50-60 years, 1. In 50 cases of tuberculosis of the testicle observed by Kocher the age of the patients was as follows: 3 cases less than 20 years; 18 cases from 20-30 years; 8 cases from 30-40 years; 11 cases from 40-50 years; 7 cases from 50-60 years; 3 cases from 60-70 years. In 69 cases analyzed by Simmonds the following was the result: 2 children from 1½-17 years; 15 patients from 20-30 years; 15 patients from 30-40 years; 16 patients from 40-50 years; 6 patients from 50-60 years; 5 patients from 60-70 years; 1 patient from 70-80 years.

It appears that tuberculosis of the testicle is most prevalent during the most active period of sexual function—that is, in patients from twenty to thirty years of age. Kocher remarks that the disease frequently attacks young men soon after marriage. This corresponds with my own personal observations. Jullien[1] reports 17 cases of tuberculosis of

---

[1] " De la tuberculose testiculaire." *Arch. Gén.*, 1890.

the testicle in children in the service of Lannelongue. Of these cases, 6 were less than two years of age; the remaining patients were from two to thirteen years old. Heredity could be traced in 4 of 10 cases. The disease often appears in the acute form, or at least with symptoms of subacute orchitis. In 12 of these cases the spermatic cord was affected. In 4 cases the affection was complicated by hydrocele of the tunica vaginalis, and in 1 case the prostate and the vesiculæ seminales were implicated. In children there is little, if any, tendency to extension of the disease to other organs. The affected organ is generally destroyed either by ulceration or absorption, a complete *restitutio ad integrum* being rare. In spontaneous cases the organ atrophies and is usually covered by a pale adherent scar. Hutinel and Deschamps[2] maintain that tuberculosis of the testicle in children is by no means infrequent. In children the disease occurs most frequently in the form of an acute infiltration. It is seldom a primary affection, but forms a part of a general diffuse tubercular process. The peritoneum especially is frequently involved. The chronic form is often overlooked because it occurs as a chronic painless induration. Otherwise the disease resembles the same affection in the adult, resulting in caseation and abscess-formation. It is only in such cases that the authors favor an operation. The results of castration in children are not encouraging. These authors are more

---

[1] "Étude sur la tuberculose des testicule les enfants." *Arch. Gén.*, 1891, p. 257.

inclined to conservative treatment by local applications and internal medication.

Rintelen[1] collected 25 cases of double tuberculosis of the testicle, besides 6 cases which he saw in Rosenberger's clinic. In 15 of these cases sufficiently accurate data could be obtained in reference to the course of the disease. The right testicle was affected first 10 times; in the remaining cases the disease commenced in the left testicle. In only 1 of these cases was the patient less than ten years of age. Three of the patients were from twenty to thirty years old; 6 were from thirty to forty, and 14 were more than forty years old; of the last number, most of the patients were between fifty and sixty years of age.

Reclus[2] is of the opinion that tuberculosis of the testicle can exist as a local affection without any tendency to dissemination, local or general. Clinical observation has shown that in about one-half of all cases of testicular tuberculosis pulmonary phthisis is absent, while autopsies show that the lungs are not implicated in about one-third of all the cases. In about 50 per cent. of all the cases the disease is met with in persons before the age of puberty, while it is found in about $2\frac{1}{2}$ per cent. of all patients suffering from pulmonary tuberculosis in persons over fifteen years of age.

As exciting causes most authors enumerate traumatism and chronic gonorrheal inflammation in the

---

[1] "Ueber Hodentuberculose mit Berücksichtigung des doppelseitigen Auftretens derselben." Inaugural Dissertation, Würzburg, 1888.
[2] " Du tuberculose du testicule et de l'orchite tuberculeuse." Thèse de Paris, 1876.

posterior portions of the urethra and the epididymis. Cryptorchism is mentioned by Nepveau and Kocher as one of the most potent of the exciting causes. Gonorrheal epididymitis is mentioned frequently as a precursor, and often imparts to the tubercular process a very malignant type. Such a case is reported by Birch-Hirschfeld.[1] A soldier twenty-four years of age and in perfect health contracted gonorrhea which led to acute epididymitis. In the course of eight days he died of miliary tuberculosis. Miliary tuberculosis was found in the peritoneum, especially well marked at the internal inguinal ring on the side of the affected testicle; miliary tuberculosis of the pleuræ, lungs, meninges, liver, spleen, and kidneys also existed; the epididymis was transformed into a cheesy mass. In the testicle itself numerous intercanalicular miliary tubercles were found, and a few cheesy nodules the size of a pea. According to Salleron,[2] of 51 cases of tuberculosis of the testicle, 4 times the testicle was affected, 37 times one epididymis, 10 times both epididymes. With the exception of tuberculosis of the remaining genito-urinary organs, he saw tuberculosis of other organs only in 1 case; only in 2 cases did the disease prove fatal. Of 47 cases, 36 were from twenty to thirty years of age. It will be seen from the statistics quoted that while no age is entirely immune to tuberculosis of the epididymis and testicle, the disease occurs most frequently in men from twenty to thirty years of age, at a time when the sexual organs are in a state of very great

---

[1] *Archiv f. Heilkunde*, 1871, H. 6.
[2] *Arch. Gén. de Méd.*, July and Aug. 1869.

physiological activity. As exciting causes figure most prominently gonorrheal epididymitis and traumatism.

Pathology.—Sir Astley Cooper, in his classical work *Observations on the Structure and Diseases of the Testis*, London, 1841, p. 162, gives the following pathological description of what he called "scrofulous inflammation of the testis:" "Upon examining the epididymis and testis when affected with this disease, I have found a yellow spot in the former, surrounded with a zone of inflammation. When the spot ulcerates in the center, the matter which it contains is not pure pus, but is composed of fibrin and serum with a slight yellow tinge. I have seen such spots in the globus minor, but more frequently seated in the globus major of the epididymis. In the testis there are generally several similar yellow spots accompanied by the same inflammatory zone, and yellow streaks are also found amidst the tubuli. Scrofulous abscesses in the testes are sometimes accompanied by a granular swelling like that which exists in the simple chronic disease."

That the tubercular nature of the majority of cases of chronic inflammation of the testicle has been admitted only for a comparatively short time is evident from a paper written in 1870 by B. Beck.[1] He insisted that it would become necessary to separate from what had formerly been included under the head of scrofulous affections of the testicle some cases which were of a tubercular nature. The tubercular form of orchitis, he claimed, seldom if ever

---

[1] "Zur käsigen Infiltration und multiplen Abscessbildung des Hodens." *Deutsche Klinik*, No. 1 und 2, 1870.

existed as an isolated affection; the complicating tubercular affections in other organs he regarded as an important diagnostic aid in differentiating between the tubercular and scrofulous forms of inflammation of the testicle. Miliary nodules of the testicle he saw only once, in the case of a child who died of miliary tuberculosis.

The naked-eye morbid appearances of tubercular epididymitis and orchitis are fairly well understood by the mass of the profession. We find here, as elsewhere, the same retrograde metamorphoses of the tubercular product—coagulation-necrosis, caseation (and in the majority of cases liquefaction of the caseous material), only exceptionally spontaneous arrest of the disease by sclerosis or calcification of the degenerated products of the tubercular inflammation.

Some doubt still remains in reference to the primary starting-point of the inflammatory process, the histological structure of the tubercle-tissue, and the manner of local dissemination of the disease.

In primary tuberculosis of the testicle, cavities the size of a hazel-nut to that of a walnut are found. Often only one such caseous cavity is met with, and the number of such cavities is always limited. If the whole organ is affected, it forms a hard body with either homogeneous cut surface or oftener with irregular figures upon the section. In other cases only one nodule in the otherwise healthy testicle is found, very much in the same way as in gummatous disease of the organ. Softening and abscess-formation follow in rapid succession. (See Fig. 5.)

Tubercular infection from the testicle may extend to the retroperitoneal lymphatic glands without implication of the epididymis and vas deferens. Sommer[1] relates the case of a man thirty-six years of age, the subject of tuberculosis of both testicles, who

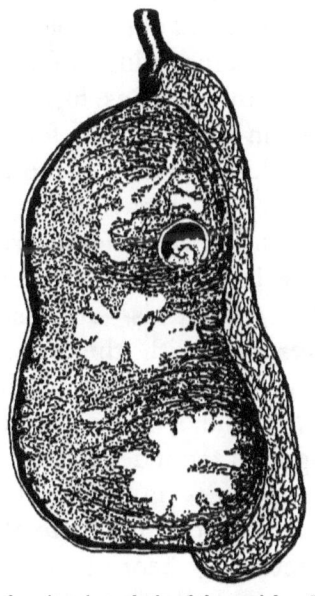

FIG. 5.—Primary chronic tuberculosis of the testicle. The testicle is much enlarged and contains a number of irregularly outlined caseous foci; the epididymis is intact, but is elongated, flattened, and stretched over the enlarged testicle.

died fifteen months from the time the first symptoms appeared. Two months before death pulmonary tuberculosis developed. In both testicles there were found from six to eight caseous foci the size of a pea, and numerous small gray nodules. There was also extensive tubercular disease of the mesenteric

[1] *Med. Zeitung in Preussen*, 1836.

glands. The tubercular disease of the genital organs was limited to the testicles. It is reasonable to assume that the infection extended from the testicles to the retroperitoneal glands, and that the disease extended from here to the lungs.

Reclus[1] makes a sharp distinction between tuberculosis of the testicle and of the epididymis. According to this author, tuberculosis of the latter nearly always occurs in the caseous form. Occasionally it presents itself in a reticulated form composed of the dilated convoluted tubules of the epididymis, the caseous contents of which have fallen out in the sections. Very often the vas deferens is affected, but, according to Reclus, never further than 5–6 centimeters from the epididymis—an observation which does not correspond with the experience of the author, who has repeatedly found the entire cord involved from the epididymis to the seminal vesicle.

Reclus recognizes the independent localization of the tubercular process in different parts of the genital tract, and places little weight on the descending or ascending theory of the infective process, as he has repeatedly found tubercular nodules of the same age and size in the epididymis and prostate. In 79 cases he examined *in vivo* he found the disease unilateral in 21, while the seminal vesicles were invariably affected on both sides. That this observation is not entirely reliable becomes very evident from Fenger's case, previously related.

It is generally conceded that the epididymis is much more frequently affected than the testicle,

---

[1] "Du tubercule du testicule et de l'orchite tuberculeuse." *Thèse de Paris*, 1876.

because reliance has been placed mainly upon the results of clinical examination. In 34 autopsies Reclus found the epididymis affected singly in only 7 instances, 27 times simultaneously with the testicle. Tuberculosis of the testicle without a similar affection of the epididymis is an exception, as he has found only 3 such cases in literature. In the testicle the tubercular process is met with either in the form of caseous foci or of miliary infiltration, or both forms are combined. The arrangement of the tubercles is usually symmetrical, corresponding to the division of the seminal tubules. The nodules are usually found in the periphery of the organ, while the caseous foci are centrally located. Fibrous tubercles which pursue a chronic course are also found in the testicle. Microscopical examinations have satisfied Reclus that the miliary form cannot be separated so easily from the caseous variety as taught by Virchow. With Malassez he locates the primary nodules in the walls of the seminal tubules, and not, as was done by Tizzoni, Gaule, and Steiner, in the intercanalicular connective tissue. He was able to remove the nodules when he resected portions of the seminal ducts, showing the connection of the nodules with the ducts. He believes that the process begins in the endothelial sheath which surrounds the tubules, which envelope, according to Ranvier, constitutes a continuous sheath of all tubules, and he maintains that the interior of the lumen is affected secondarily.

During the progress of the disease "granulations composées" are formed, which can only be isolated with portions of several tubuli seminiferi. While this histological process is regarded by the author as characteristic of testicular tuberculosis in the epididymis, the process begins in the subepithelial elements of the tubules.

The microscopical appearance of tubercle-tissue in the parenchyma of the testicle is the same whether the disease occurs as a primary affection or in consequence of extension from the epididymis. In the primary form the foci are few and large, varying in size from a hazel-nut to that of a walnut. During the early stages of the disease the nodules are much firmer than the surrounding normal parenchyma of the organ. The mass is surrounded by a vascular zone. Central caseation, softening, abscess-formation, and perforation often follow in quite rapid succession. If the disease of the testicle is secondary to tuberculosis of the epididymis, the nodules are more numerous and the disease presents more the appearance of an infiltration. The tubercles are formed between the seminal tubules, which are separated from each other by the tubercular product. The interstitial connective tissue as well as the adventitia is infiltrated with small round cells. The vascularity of the affected part is at first increased, but as the nodules increase in size the vessels disappear in the center, and with them the tubules. According to Rindfleisch, only the adventitia disappears, while the propria becomes edematous, but remains and can be identified in the cheesy product for a long time.

Curling,[1] in the later editions of his work, has

---

[1] *On Diseases of the Testis*, Phila., 1878, p. 335.

expressed the opinion that the disease is originally developed within the tubules of the testicle, and the results of microscopical examinations have induced him to adopt this view. He says: "Anatomical considerations, indeed, support the opinion that abnormal nutrition in the cellular contents of the tubes induces the formation of miliary tubercles in their walls without at all negativing the development of tubercle in the intertubular tissue as seen by Virchow, or in the adventitia of the blood-vessels as observed by Nepveau. Indeed, the discrepant views upon the matter may be explained by assuming that different observers have regarded what has been found in particular cases as the result of some general law. With reference to this, the suggestion of Klebs is valuable. Admitting that in acute miliary tuberculosis, where the dissemination of the virus is effected by the vascular system, the blood-vessels and their surroundings are the seat of the tubercles, he has seen preparations by Langhans where the tubercles were in the interior of the tubules; and Klebs adds that 'it would be very desirable to ascertain whether this was uniformly the case in the so frequent extension of tuberculosis from the older nodules in the epididymis to the body of the testis.'"

Salleron[1] as early as 1869 observed in his military practice 51 cases of tuberculosis of the testicle. The testicle itself was affected only 4 times, the epididymis on one side 37 times, on both sides 19 times. Only in 1 case was he able to ascertain *intra*

---

[1] "Mémoire sur l'affection tuberculeuse des organes génitaux de l'homme." *Arch. Gén. de Méd.*, July and August, 1869.

vitam the existence of pulmonary tuberculosis—a fact which is in opposition to the experience of Curling and Louis, but which he supported by 9 accurately reported cases and 2 autopsies. With the exception of the vas deferens and the seminal vesicles, he found the remaining portions of the urogenital tract free from tuberculosis.

Years ago, Friedländer[1] found miliary tubercles in the testicle in cases in which no other organ was found affected. Nepveau[2] found miliary tubercles upon the walls of blood-vessels in diseased testicles complicating secondary renal and primary pulmonary tuberculosis.

Rindfleisch calls attention to the unusual size of tubercles in tuberculosis of the testicles; the tubercles, according to his observations, vary from the size of a pin's head to that of a walnut. In the testicle the first infiltration usually shows itself as fibroid tubercle, as a light yellow or grayish-white hard and tough nodule in the parenchyma of the organ. In miliary granulations the tubercle appears under the microscope as described by Langhans—in the center a giant cell, around this a zone of epithelioid cells, and in the periphery a small round-celled infiltration, the cellular elements embedded in a reticulum of connective tissue. (See Fig. 6.) Kocher[3] is also of the opinion that the primary starting-point of the disease in the epididymis is in the intertubular connective tissue. The contents of the tubules are in-

---

[1] *Sammlung klinischer Vorträge*, 1873.
[2] *Contribution à l'Étude les Tumeurs du Testicule*, Paris, 1872.
[3] *Krankheiten des Hodens, Nebenhodens und Samenstranges.* Pitha u. Billroth, B. iii. p. 3, B. p. 273.

FIG. 6.—Histological structure of tubercle in the testicle; × 250: *a*, giant cell; *b*, epithelioid cells with leucocytic infiltration; *c*, tubule.

FIG. 7.—Histological structure of tubercle of the epididymis; × 160: *a*, tubules; *b*, tubercle with giant cells; *c*, edge of mucous membrane; *d*, smooth muscle-fibers; *e*, tunica vaginalis.

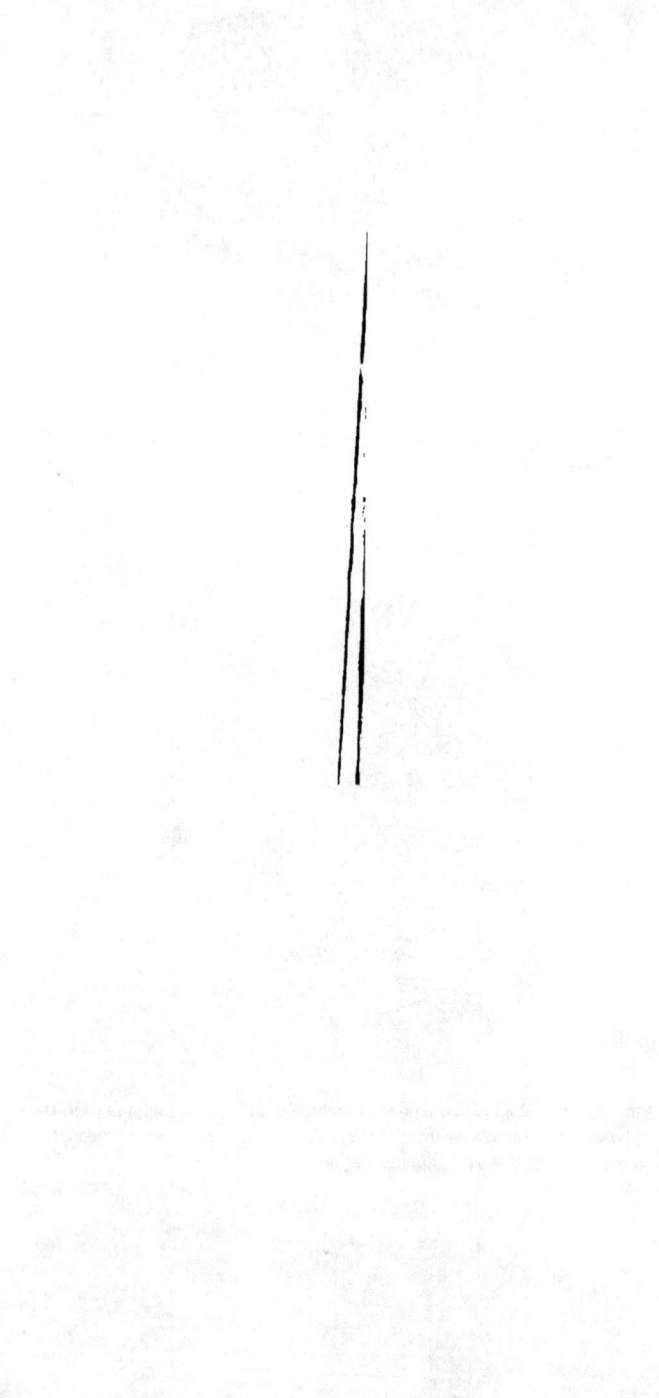

creased by proliferation of the pre-existing tissues of their walls. The infection begins most frequently in the globus major, extending thence to the remaining parts until the entire organ is transformed into a

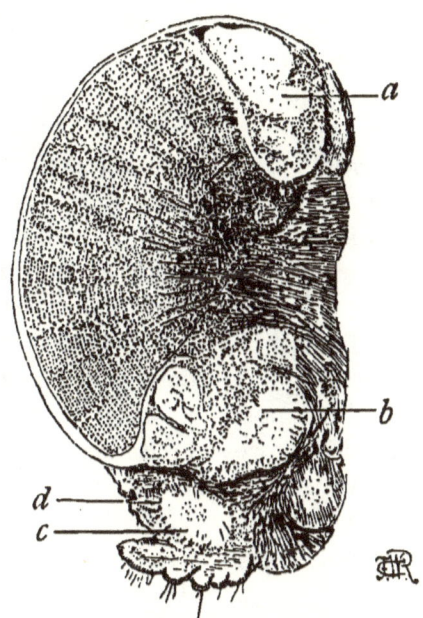

FIG. 8.—Tuberculosis of the epididymis extending to the testicle by continuity of the tubercular process.

hard nodular mass (see Fig. 8). Cheesy degeneration of the contents of the tubules results in destruction of their walls and coalescence with the intertubular products, when a tubercular abscess forms, which frequently ruptures spontaneously.

During the early stage of the disease the cut surface has a grayish glistening appearance and is traversed by yellow stripes marking the location and course of the seminal tubules. This pathological condition precedes more extensive caseation. The tubules become dilated by accumulation of caseous material, which in sections falls out. As the disease advances the tubular spaces coalesce and form caseous cavities of variable size. The tissues exposed to the pathogenic action of the bacillus tuberculosis—notably, the connective tissue, the endothelial cells of the capillaries, and the epithelial cells of the seminal tubules—undergo karyokinetic changes that result in the formation of epithelioid cells, the first histological elements of the primitive tubercle-tissue. Later, migration of leucocytes increases the tubercular product. Giant cells form where the tubercle bacilli are present in limited number. Whether these cells result from tissue-proliferation or from confluence has not as yet been positively ascertained; the former supposition is, however, the more probable. As the disease advances the epithelioid cells are often overshadowed by the leucocytes, giving to the tissue the appearance of what was formerly demonstrated as "lymphoid tubercle." Giant cells are usually present. The small giant cells are usually round; the large ones send out numerous prolongations. The reticulum is better developed in the tubercles of the epididymis and testicle than in tubercular affections of any other organ.

Isolated epithelioid cells are usually found in the meshes of the delicate reticulum. Some of the spaces formed by the reticulum are empty. The reticulum is formed of new tissue which is probably derived from the epithelioid cells.

The cellular elements of the membrana propria take an active and important part in the histogenesis of tuberculosis of the epididymis and the testicle (see Fig. 7).

According to Weigert, the giant cells are produced by cells infected with tubercle bacilli, which produce a partial necrosis, and this change again prevents cell-segmentation; the latter in turn is followed by accumulation of nuclei which arrange themselves in the periphery of the cell-protoplasm.

Gaule[1] regards tuberculosis of the testicle as a catarrhal inflammation of the epithelial lining of the seminal tubules that leads to stagnation of the secretions and to caseous degeneration of the inflammatory product, which, owing to the thinness of the walls of the tubules, is prone to undergo ulceration. According to this author, the process begins in the epididymis, and later extends to the testicle, when it assumes another character, as the intertubular connective tissue soon takes part in the process of tissue-proliferation. In the epididymis the process is prone to remain circumscribed, and is favorable to the development of a fibrous nodule limited to the interior of a single tubule. While the contents of the nodule may undergo caseation, the existing irritation extends to the adjacent intertubular tissue, and gives rise here to fibrous products frequently in the vicinity of the septa. Later, adjacent tubules are included in the process and undergo similar changes, constituting the condition described by Reclus as *granulations composées*.

The local and general dissemination of tuberculosis of the testicle and epididymis is well shown

---

[1] "Anatomische Untersuchungen über Hodentuberculose." *Virchow's Archiv*, 1877, B. lxix. pp. 64, 213.

by the observations of Guyon,[1] who found in 28 post-mortems on persons the subject of tubercular disease of the testicle, that the lungs were affected in only 11 cases. In 222 clinical observations on patients suffering from urogenital tuberculosis that he examined during a period of twenty-five years, 40 were cases of isolated genital tuberculosis, 74 were cases of tuberculosis of the urinary organs, and only 108 were cases of combined urogenital tuberculosis. Of 42 additional post-mortems, 1 was a case of isolated genital tuberculosis, 14 were cases of combined tuberculosis of the urinary organs, and 27 were cases of combined urogenital tuberculosis. Seldom is the testicle or the epididymis the only part affected. Guyon believes, with Lancereaux, that the tubercular process begins very frequently in the vesiculæ seminales. In 13 cases all the genital organs were affected. In 127 clinical cases in persons before the age of puberty suffering from urogenital tuberculosis, the prostate was affected 56 times, the prostate and the seminal vesicles 11 times, the epididymis alone 2 times, and all the genital organs 58 times. Among these there were 2 cases in which it could not be positively ascertained that the prostate was involved, and the same uncertainty existed in reference to the seminal vesicles in 6 cases. According to these statistics, the course of the disease in men is therefore more in the direction from within outward than from without inward—a fact which Guyon advances as a warning against the too frequent performance of castration.

[1] "Clinique des Maladies des voies urinaires à Necker, La Castration pour Sarcocele." *Ann. des Mal. des org. Gén.-urin.*, 1891, p. 445.

Not infrequently the tubercular affection extends from the epididymis or the testicle to the tunica vaginalis. Simmonds[1] made a careful examination of 8 tubercular testicles obtained from 6 patients operated upon in the clinic at Kiel, and of postmortem specimens obtained from the hospitals at Hamburg—in all, 12 testicles. In 8 of these specimens tubercles were found in the tunica vaginalis, in 1 case the testicle was atrophied, while in 3 specimens the tubercular process was not far advanced. These cases appear to prove the incorrectness of the statement made by Klebs, that the tunica vaginalis is never affected; at the same time they call the attention of the surgeon to the necessity of examining the tunica vaginalis carefully and of subjecting it to operative treatment if the disease has extended to it in cases of tuberculosis of the testicle treated by castration.

Symptoms and Diagnosis.—The disease presents itself clinically in the form of a caseous inflammation with ulcerative suppurative catarrh. Tuberculosis of the testicle and epididymis is a very insidious disease. It is often overlooked by the patient for a long time, and frequently is erroneously diagnosticated by the physician. The disease of the testicle before it is detected is often preceded by a slight urethral discharge, probably mistaken frequently for one of venereal origin. In the absence of tuberculosis of other portions of the genital organs and the urinary tract, the disease usually begins in the globus major of the epididymis, much less frequently

[1] "Ueber Tuberculose der Scheidenhaut des Hodens." *Deutsche Zeitschrift f. Chirurgie*, B. xviii. p. 157

in the body or the opposite pole of the organ, as a hard, almost painless swelling. During the progress of the disease additional nodules form, and very frequently the patient's attention is attracted by a more rapidly increasing swelling—a complicating hydrocele. The absence of any well-marked symptoms during the incipiency of the disease is the reason why such cases come so rarely under the care of a surgeon at this time. From the globus major the disease spreads to the body and the globus minor, and then extends along the vas deferens.

After an elaborate description of the signs and symptoms of scrofula, Sir Astley Cooper[1] gives a very vivid clinical picture of what was then considered as scrofula of the testicle, in the following language: "One of the testicles, even in very young children, sometimes becomes enlarged and very hard, but without pain or any other inconvenience, and the disease is accidentally discovered by the parents or by a servant. In this state of indolent increase it remains for many weeks, months, or years, and then, under improvement of the general health, the enlargement subsides and the gland resumes its former state. More frequently it enlarges at the age of puberty and from that period to twenty years, and not infrequently this disease appears in both testes, marked by the same hardness and by such absence of suffering that the person does not for a length of time seek any medical aid respecting it. The part is free from pain as well as tenderness. The scrotum is unaffected, its veins are

---

[1] *Op. cit.*, p. 160.

not enlarged, and, but from its bulk, the patient suffers no inconvenience; but even in children, although more frequently at puberty, the inflammation sometimes proceeds to suppuration. This generally occurs within the globus major of the epididymis, but I have known abscesses to form first in the cauda, or small extremity of that organ. The body of the testicle but rarely suppurates until after the epididymis has ulcerated, when the testis becomes affected and the scrotum puts on a livid hue. Ulceration next ensues, indicating the presence of an abscess, which discharges ill-formed pus and some semen, if after the age; the opening under these circumstances is extremely difficult to heal, continuing for months and even for years before it closes. In some persons one abscess after another forms and discharges, and when one testis has suppurated, if the other has been hard, it is liable to put on the same action, discharges itself, and continues obstinately resisting all the means of treatment for a greater length of time. Ultimately the testes diminish in size, secrete but a small quantity of semen, and continue to waste until but little of them remains and their function almost entirely ceases." Astley Cooper and his contemporaries had no correct idea of the intrinsic tendency of tubercular inflammation of the testicle to extend to the remaining organs of the genital tract and the urinary apparatus. The essential clinical features of this disease have been elaborated since their time.

Reclus[1] is of the opinion that chronic orchitis and

---

[1] *Op. cit.*

epididymitis are often confounded with tuberculosis. The abundance of interstitial connective tissue produced in the course of these affections leads constantly to progressive atrophy. These conditions are often characterized by a nodular condition of the swelling which closely resembles tuberculosis. They occur either in the course of an acute process or begin as chronic processes; usually, however, the swelling is much more marked than in tuberculosis. If the testicle alone is affected, it finally is reduced in size to that of a bean in front of the normal epididymis. Kocher regards as the most characteristic symptoms of tuberculosis of the testicle and epididymis rapid development of swelling, early softening of the inflammatory product, and the absence of acute subjective symptoms. According to this writer, the maximum swelling is often reached in eight days or at least in a few weeks; the swelling soon softens and perforates, resulting in the formation of a fistula which may continue for years, the swelling remaining stationary. Kocher's observations in regard to the acuity of the local development do not correspond with the result of my own experience. Such a course is an exceptional one. Barling[1] describes a case of double galloping tuberculosis of the testicle, an affection to which Duplay called the attention of the profession as early as 1876. The patient, who was thirty-one years of age, five hours after a severe external injury noticed quite a large swelling of one of the testicles. One of the interesting features in this

[1] "Double Acute Tubercular Disease of the Testis." *Lancet*, London, April 9, 1892.

case was the fact that the epididymis was free, the disease being limited to the testicle. In a short time the opposite organ became similarly affected. Examination of the testicles after castration showed in each of them cheesy cavities—a large one in the right testicle and a number of small ones in the left. The immediate vicinity of the cavities was occupied by dense and greatly increased connective tissue. In one place, in a cavity, tubercle bacilli were found, but the most patient search did not result in finding giant cells. The whole process was characterized by an infiltration with granulation-cells and epithelioid cells, combined with catarrh of the seminal tubules followed by caseation. I can readily understand, in the event of tuberculosis developing in a testicle or an epididymis the seat of an injury or an antecedent inflammatory disease, that it might in rare cases pursue such an acute course; but in the majority of cases the affection is noted for its insidiousness and chronicity. The patient's attention is usually first attracted by slight pain or discomfort and tenderness in some part of the gland, generally the epididymis, which, on examination, is found to be somewhat enlarged, prominent, nodular, and indurated. The state of the testicle is often marked by circumscribed effusion of fluid in the tunica vaginalis, the surfaces of this membrane being partially adherent. The disease often remains stationary for months, or even for a year or more. During the course of the disease one of the prominences begins to increase in size, so as to be observed externally and to feel more painful and tender than the other nodules; the skin over it becomes adherent, changes to a livid color, and ulcerates, when the softened cheesy material is evacuated. The abscess-formation is generally followed by a fistula which communicates with the primary tubercular focus. Similar changes may take place in other parts of the testicle, resulting in two or more sinuses leading into the interior of the gland. If all the tubercular material is eliminated in this manner, the sinuses after a long time may heal, leaving the testicle usually in an atrophied condition. The disease may remain limited to one testicle, or after months or years may make its appearance in the opposite organ. As a rule, the vas deferens is early affected, the infection often extending its entire length in a short time. In primary tuberculosis of both testicles both vasa deferentia may remain intact. The tunica vaginalis is usually implicated in the form of an adhesive periorchitis. As this affection is not attended by any well-marked clinical symptoms, its existence can be surmised by the appearance of hydrocele. The hydrocele, usually of a serous type, may attain a considerable size, but more frequently it is circumscribed. In rare cases the vaginalitis is of a suppurative type, and then makes its appearance under the clinical picture of an acute abscess. Tubercular affections of other organs precede or occur simultaneously with the tubercular orchitis. Next to the vas deferens, the disease extends most frequently to the seminal vesicles; next, the lateral lobes of the prostate are most frequently implicated. From here the disease is prone to involve the urinary organs—first the bladder, next the ureter, and finally the kidney. In the differential diagnosis must be considered acute and chronic inflammations of another type and syphilitic affections. In tuberculosis the swelling in the epididymis is usually larger than in other forms of inflammation. Tenderness and pain, conspicuous symptoms in gonorrheal epididymitis, are absent or slight in tuberculosis. If any doubt exists in the diagnosis between gumma and tuberculosis, the patient should be given the benefit of the doubt and should be subjected for a sufficient length of time to an active antisyphilitic treatment. The tubercular inflammation is clinically characterized by periodical exacerbations. Injection of tuberculin may prove of diagnostic value. Examination of the remaining genitourinary organs for tuberculosis, as well as the more distant organs in which this disease is liable to appear, is absolutely necessary, and will often clear up a doubtful diagnosis. In tuberculosis the indurated vas deferens is not very tender to pressure and is usually nodular, while funiculitis caused by other forms of infection is attended by pain and tenderness.

Treatment.—The rational treatment of tuberculosis of the testicle and epididymis must necessarily depend on the location, stage, and extent of the disease and the existence or absence of complications. In miliary tuberculosis involving the organs on both sides the treatment must be supporting and expectant, as in such cases a speedy fatal termination is inevitable from the primary pulmonary affections or from diffuse miliary tuberculosis. In tuberculosis involving other parts of the urogenital organs the treatment must be directed accordingly.

Tuberculosis implicating the organs on both sides simultaneously or in succession is almost positive proof of the existence of an older tubercular process in some other organ, or the extension of the tubercular process along the genital tract beyond the reach of successful surgical treatment. The cases best adapted for successful local treatment are those in which the disease appears as a chronic affection and is limited to the organ on one side, with limited or no extension along the vas deferens. Tuberculosis of the epididymis leads to impotence if both organs are affected. If the disease is limited to one side, function may remain unimpaired. Castration must therefore be regarded as the normal procedure in cases of uncomplicated unilateral tubercular epididymitis. This statement becomes more apparent and forcible when we consider that reinfection can always occur if the diseased organ is permitted to remain. As positive contraindications to castration must be considered the following: 1. Extension of the tubercular process to parts not within reach of a radical operation; 2. Tuberculosis of both testicles, as the second testicle is usually involved by extension of the infection from one to the other through the vasa deferentia; 3. Tuberculosis of important adjacent or distant organs.

Castration may become necessary as a palliative operation for the relief of symptoms in cases where the disease has resulted in the formation of tubercular abscesses and suppurating sinuses. A great diversity of opinion still prevails among surgeons regarding the value of castration in the treatment of tuberculosis of the testicle and epididymis. The results following this operation appear to have been quite at variance in the hands of different operators.

Terillon[1] favors castration, and advises that the operation should be performed before abscesses have formed, as by removing the source of suppuration the general condition of the patient is improved. Pulmonary tuberculosis he regards as an absolute contraindication to castration. Simultaneous tuberculosis of the prostate and the seminal vesicles may or may not be regarded in the same light, according to circumstances.

Richet is nearly of the same opinion in reference to the indications for castration, only he opposes the operation more decidedly when extension of the disease to other parts of the genital tract has occurred.

Stenger[2] gives the result of 13 cases of castration for tuberculosis of the testicle that occurred in the Royal Clinic at Berlin from 1883 to 1889: 7 of the patients remained free from local recurrence or evidences of infection of any other organ at the time the report was made; in 3 of the cases tuberculosis of other organs existed at the time the operation was performed; and in the last 3 cases the final result could not be ascertained. In 7 other cases the tubercular product was removed by vigorous use of the sharp spoon—in 5 cases with a satisfactory result, and in 2 cases pulmonary infiltration existed at the time the operation was performed, and progressed uninfluenced by the operation. From a study of these cases he came to the conclusion that

---

[1] "Essai critique sur le traitement de la tuberculisation du testicule." *Bull. Gén. Thérap.*, 1882, p. 140.
[2] Inaugural Dissertation, Berlin, 1889.

the timely removal of the diseased organ or a thorough local operation is instrumental in preventing general infection.

Finkh[1] has ascertained the ultimate results in 29 cases of tuberculosis of the testicle treated by castration in Bruns' clinic at Tübingen. Of these cases, the right testicle was affected 12 times, the left 8 times, and 9 times both testicles were diseased. At the time the report was made, in 1886, 14 of the cases were living and free from relapse or tubercular disease of any other organ. Among these cases were 9 in which only 1 testicle was removed, and 5 double castrations. Of the first group, one to twenty-nine years had elapsed since operation; of the second, one to ten years. Of those that died, 8 had succumbed to non-tubercular affections. Of these, 5 were single and 3 double castrations, the former with a period of immunity varying from nine months to twenty-two years, the latter from five to thirty years. To these must be added a case of double castration in which death occurred twelve days after operation, from exhaustion. Six cases died of tuberculosis—all single castrations—and of these cases, only in one instance did the disease extend to other parts of the genital tract. In 1 case life was prolonged for four years, and against the 6 cases that died of tuberculosis stood 13 that remained well after five years and more. The infected pelvic portion of the cord furnishes, according to this author, no contraindication to an operation, as in 7 cases in which this condition was found the

---

[1] "Ueber die Endresultate der Castration bei Hodentuberculose." *Beiträge zur klin. Chirurgie*, B. ii. p. 407.

result of the operation proved satisfactory. These results are certainly more favorable than those obtained by the average surgeon. In my own cases I have frequently observed relapse in unilateral castration, and in 2 cases of double castration the disease, after a year or two, attacked the seminal vesicles, prostate, and bladder, and death finally resulted from tubercular pyelonephritis.

It seems to me that the cases are exceptional in which double castration is justifiable. Castration appears to have yielded satisfactory results in cases in which the disease was unilateral and the testicle was removed before the disease extended to other parts of the genital organs. A number of cases in which the disease was unilateral and was complicated by tubercular vesiculitis, and which were treated by castration and excision of the seminal vesicles and even a portion of the prostate, resulted favorably. In the removal of a tubercular testicle it should be taken for granted that the vas deferens is infected, and as much of this structure as possible should be excised.

Kocher does not recommend that the cord should be followed as high up as possible, but advises that it should be cut at a point above where it shows decided pathological changes. He places great stress upon the importance of cauterizing strongly the lumen of the cut end with the fine needle-point of the Paquelin cautery or with pure carbolic acid, and upon burying the stump with a few catgut sutures for the purpose of guarding against tubercular infection of the wound.

Conservative operations of different kinds have

been made for a long time in the treatment of the so-called scrofulous testicle. Malgaigne excised caseous nodules with the bistoury; Delpech, Boyer, Velpeau, Bonnet, and Bouisson gave preference to chemical caustics such as chloride of zinc, caustic potassa, and Vienna paste; later, Verneuil recommended the actual cautery, which was replaced later by the Paquelin cautery. The use of the thermocautery is strongly recommended by Forgue and Reclus[1] in cases in which the cheesy depots are few or single. The same authors are in favor of iodoform-ether injections (10 : 90) when the affection is more diffuse. A few drops of this solution are injected into each nodule and the little punctures sealed with iodoform collodion. In cases not amenable to conservative treatment they recommend without hesitation castration. I have used for some time, during the early stages of tubercular epididymitis, parenchymatous injections of iodoform-glycerin emulsion with the most satisfactory results. With a small trocar the epididymis is penetrated from end to end, and the injection is made slowly as the cannula is withdrawn. Under moderate pressure from 1 to 1½ drams of the emulsion can be injected, thus permeating the affected tissues with the antibacillary agent. The injection should be repeated every week or two. If the disease is complicated by hydrocele, the fluid should be evacuated and a small quantity of the emulsion injected. The pain following this treatment lasts only for a short time. For a few days the swelling increases and tenderness is more marked.

---

[1] *Traité de Thérapeutique Chirurgicale*, T. 2, p. 912.

The increased tissue-proliferation that is excited by the action of the iodoform is a potent element in arresting the extension of the disease and in preventing further degenerative changes in the tubercular tissue.

Terillon[1] has given evidement and iodoform-gauze tamponade a fair trial in the treatment of tuberculosis of the epididymis advanced to the stage of abscess- and fistula-formation, but on the whole he gives castration the preference, as he believes evidement, cauterization, and drainage, as a rule, yield only temporary beneficial results.

Keyes[2] removed by means of the curette an entire tubercular epididymis. A section of the spermatic cord being then found to be extensively ulcerated, 1¼ inches of that was removed. The function of the testicle had already been destroyed by cheesy foci along its course, and the patient knew that he was practically castrated before the operation was undertaken, but he was much more pleased with the result than if the testicle had been removed. The wound healed rapidly by first intention, the relief to the patient being complete.

Villeneuve[3] prefers thorough cauterization with the Paquelin cautery to castration in the treatment of tuberculosis of the testicle. He cites cases, and defends his position by what appears to be plausible reasoning. Kocher uses the sharp spoon followed by the application of a solution of chlorid of zinc

---

[1] "De l'intervention chirurgicale dans la tuberculose testiculaire." *Progrès Méd.*, No. 3, 1886.
[2] *Annual of the Universal Medical Sciences*, 1892, vol. iii. p. 3 F.
[3] *Marseille-Médicale*, July 30, 1889.

(1:4) or by repeated applications of the strong tincture of iodin when the disease has advanced to the formation of abscesses and fistulæ. Lannelongue speaks highly of the use of a solution of chlorid of zinc as a parenchymatous injection before suppuration has occurred. The injection is made around, and not into, the tubercular infiltration.

The author has resorted to the sclerogenic method in a number of cases for the purpose of preventing the extension of the infection along the vas deferens, and with very encouraging results. About 15 drops of a 10 per cent. solution of chlorid of zinc are injected at four or five different points into the loose connective tissue around the cord.

Ozenne[1] reports success in 1 case of tubercular epididymitis, and favorable progress in a few others, from injections of a 10 per cent. solution of chlorid of zinc, after the method of Lannelongue. In the successful case he injected at four sittings, in seven different places of the diseased area, 2 drops at each puncture. Moderate temporary reaction followed. Some months later, one small tubercular nodule remaining, a single injection was given. After the last treatment all active signs of the disease disappeared.

From what has been said on the treatment of tuberculosis of the testicle it is evident that this subject remains an open chapter. In recent cases of primary tuberculosis parenchymatous injections of iodoform-glycerin emulsion or of chlorid of zinc should be used. If this treatment does not prove

[1] *Gazette des hôpitaux*, Feb. 23, Aug. 9, 1893.

satisfactory after a fair trial, castration should be performed before the disease extends to additional organs. In limited abscess-formation the use of the sharp spoon and the iodoform-gauze tampon may prove efficient, but if the disease resists these measures castration is strongly indicated. If the disease is bilateral, palliative treatment should take the place of a radical operation in the majority of cases. Castration is absolutely contraindicated when the tubercular affection of the testicle is complicated by tuberculosis of any important internal organ. Simultaneous tuberculosis of the prostate and the seminal vesicles does not necessarily contraindicate castration.

PART III.

TUBERCULOSIS OF THE FEMALE ORGANS OF GENERATION.

THE different forms of surgical tuberculosis have received renewed attention since the epoch-making clinical researches of Villemin and the laborious pathological and histological investigations of Virchow, but more particularly since the essential cause of tuberculosis was discovered by Koch. Surgeons were more prompt in adopting and putting into practice the modern views pertaining to the nature of tuberculosis than the gynecologists, hence the clinical history and pathological conditions of the male genital organs are much better understood and more familiar to the general practitioner than the same affections of the opposite sex. The clinical material illustrating tuberculosis of the female generative organs is scattered, and with few exceptions is very imperfectly elaborated, and there can be but little doubt that mistakes in diagnosis are of frequent occurrence to-day, as the distinction between tuberculosis, carcinoma, and syphilitic lesions, even when the affection involves the skin, is by no means always an easy task, and the difficulties encountered in making a positive diagnosis when the disease affects any part of the female genital tract are much greater than in the case of the male. I am sure that many ulcers of the vulva, vagina, and the vaginal portion of the cervix uteri have been cauterized or excised under the belief that they were of a carcinomatous nature, when a careful microscopical and bacteriological examination would have revealed their tubercular nature. Many cases of so-called fungous endometritis that did not yield to local applications and the use of the sharp spoon would upon a more careful examination be found to belong to the same category. The frequent performance of salpingectomy has furnished and will continue to furnish a rich material for the study of tuberculosis of the Fallopian tubes. Laparotomy for tubercular ascites will in the course of a few years show more satisfactorily the etiological relations which exist between tubercular peritonitis and tuberculosis of the female genital organs.

History.—The fragmentary knowledge that we now possess of the pathology and clinical history of tubercular disease of the female organs of generation is the result of the labors of men who have lived and worked during the present century; indeed, very little was known concerning this subject before 1850. According to Williams,[1] the first case of genital tuberculosis in the female was described in 1744 by Morgagni, who, at the autopsy on a girl of twenty years, dead of tubercular peritonitis, found both the tubes and the uterus filled with caseous material, and the ovaries and tubes so matted together that

[1] "Tuberculosis of the Female Generative Organs." *Johns Hopkins Hospital Reports*, vol. iii. Nos. 1, 2, and 3.

it was impossible to separate them; he did not hesitate to consider genital tuberculosis as the primary disease. Somewhat similar cases were described later by Louis and by Senn of Geneva. In 1831, Renaud described 2 cases of uterine tuberculosis in phthisical patients. Cruveilhier figures in his great work a uterus with two tubercular appendages, and a similar illustration can be found in the work on pathology by Hope, published in 1834. Kiwisch and Paulsen gave a good description of tuberculosis of the uterus. Virchow and Rokitansky regarded tuberculosis of the ovary as a pathological curiosity. Verneuil and Cohnheim were of the opinion that tuberculosis of the genital organs in both sexes is not infrequently transmitted by coition. "The removal by Marchand, in 1881, of tubercular tubes and ovaries by laparotomy, and the discovery of tubercle bacilli in the vaginal secretion by Babes, soon greatly increased the interest in the affection, both from a diagnostic and an operative standpoint. Owing to this interest, which has not yet subsided, a considerable number of articles have been written upon this subject during the last few years" (Williams).

The etiological relation of pulmonary, intestinal, genital, and peritoneal tuberculosis, and the causative relationship of genital to peritoneal tuberculosis, will have to be more accurately elicited by extended observations, post-mortem examinations, and more careful examinations of specimens removed by operation or for diagnostic purposes.

Manner of Infection.—We will find, in discussing the subject of genital tuberculosis as it occurs in the female, that there are many avenues of infection, according to the organ or part of the genital tract affected. The views of authors present and past differ greatly concerning the frequency and point of origin of genital tuberculosis. It is sometimes exceedingly difficult, and occasionally impossible, in cases of multiple tubercular lesions, to ascertain their correct chronological order. Tuberculosis of the genital organs in both sexes seldom presents itself to the surgeon as an isolated affection, and the records of pathologists who conducted the examinations after death in such cases usually show the widest dissemination of the tubercular process, involving often numerous and anatomically the most disconnected organs. As our experience increases the views of Cohnheim and Verneuil relating to the possibility of direct infection by contact with a tubercular husband or wife, and *vice versa*, will become more firmly established. There can be no question, however, that the most frequent methods of infection are through the general circulation and by extension of the tubercular processes from adjacent organs. In this connection the writer will quote the opinions of only a few of the earlier writers on this subject, reserving a more detailed account for a description of the disease as it occurs in the different parts of the genital tract.

From the cases reported and the records of postmortem examinations it becomes evident that the Fallopian tubes are most frequently affected. Spaeth[1] has collected 119 cases of tuberculosis of the female

---

[1] "Genital tuberculose beim Weibe." Dissertation, Strassburg, 1885.

generative organs, and found that in 28 cases the process involved the entire genital tract, 9 times the vagina, 10 times the uterus, 103 times the tubes, and 15 times the ovaries, either primarily or by extension from other parts of the genital tract. To these cases he added 2 new ones that came under his observation, illustrating the primary and secondary invasion of the genital organs. In the first case the disease evidently had its starting-point in the genital organs, involving the tubes, ovaries, uterine cavity, and external os. Three months after the extirpation of the uterine adnexa the patient died from the effects of the extension of the disease to more important organs. In the second case, in which the patient was the subject of pulmonary phthisis, the disease later involved the peritoneum, tubes, uterus, and the vaginal portion of the cervix uteri. The frequency with which pulmonary tuberculosis precedes, attends, or follows genital tuberculosis is well shown by the statistics gathered by Mosler,[1] who collected 47 cases of tuberculosis of the female genital organs. Out of this number the tubes were affected 34 times, the uterus nearly as often, 7 times the ovaries, 3 times the vagina. In 36 of these cases pulmonary tuberculosis existed at the same time, and 31 times peritoneal tuberculosis. In 9 of these cases the author believes the genital organs were the primary seat of the disease, and in 2 cases the genital affection was complicated by inveterate syphilis. Lebert[2] believes that primary

---

[1] "Die Tuberculose der weiblichen Genitalien." Dissertation, Breslau, 1883.

[2] "Ueber Tuberculose der weiblichen Geschlechtsorgane und über den

tuberculosis of the genital organs does occur in women, but that it is a very rare disease. If it occurs simultaneously with pulmonary tuberculosis, it presents itself in a very diffuse form. The urinary organs are much less frequently implicated in women than in men. He calls special attention to the tendency of the disease to extend along mucous surfaces by an uninterrupted continuity of surface from one part of the genital tract to others. He showed that the cervical erosions which have so often been regarded as of a tubercular nature are simply follicular ulcers of a non-tubercular character. In opposition to Lebert and many others, Gehle[1] expressed his belief that primary tuberculosis of the female organs of generation is not such a very rare disease. By observing a case of this kind and a study of 22 cases collected from different sources, he came to the following practical conclusions concerning this disease: Tuberculosis of the female genital organs manifests itself clinically in the form of a cheesy, chronic inflammation. Only in 2 cases were miliary tubercles found in the genital organs. Almost constantly the tubes are the starting-point of the disease, which thence extends to the uterus, ovaries, and peritoneum. If the lungs are secondarily affected, this complication sets in at a comparatively remote period. Gehle also called attention to the infrequency with which tuberculosis of the female genital organs extends to the

Einfluss des weiblichen Geschlechtslebens auf die Entwickelung und den Verlauf der Tuberculose." *Archiv f. Gynäkologie*, B. iv. S. 457.

[1] " Ueber primäre Tuberculose der weiblichen Geschlechts-Genitalien." *Virchow's Jahresbericht*, 1881, vol. ii. p. 535.

urinary organs as compared with the same disease in men. Although this author believes that in a fair proportion of all cases of genital tuberculosis the disease occurs as a primary local affection, he takes a gloomy view of its prognosis, and discourages operative measures in the treatment. Hegar,[1] on the other hand, maintains that genital tuberculosis may remain for a long time stationary—that is, in a latent condition—and that a spontaneous cure not seldom takes place by suppuration or encapsulation. He believes that the disease in the majority of cases is acquired by auto-infection—that is, it occurs secondarily to tuberculosis in other organs. According to this author, the genital organs are reached through the lymphatic system through the action of the ciliated epithelia of the tubes, or by the perforation of an adjacent tubercular abscess into the genital organs. The second pathway of infection is from without—by coitus or through infection during delivery or during the puerperium. The transportation of tubercle bacilli occurs in two ways: either by extension through the tissues, or from the mucosa of the tract through the lymphatics, connective tissue, and serosa. The extension of the tubercular process through the lymphatic system either to or from the genital organs is recognized as a frequent pathway of infection by most authors. Fernet[2] has even found progressive invasion of the pleura, and the production of subacute peritoneo-

---

[1] *Die Entstehung, Diagnose und chirurgische Behandlung der Genital tuberculose des Weibes.* Stuttgart, 1886.

[2] "De la Tuberculose Péritonéo-pleurale subaiguë." *Bull. de la Soc. méd. des Hôpitaux*, 1884.

pleural tuberculosis the starting-point of which was a primary lesion of the genital organs.

The infection is carried through the lymphatics, which Hegar once found to contain cheesy matter. The lymphatic communications between the pleura and the peritoneum through the diaphragm account for this method of extension of pleural tuberculosis to peritoneum and from peritoneum to the genital organs, and from the latter in a reverse direction to the peritoneum and from there to the pleura and lungs. Bryson,[1] in his most excellent article on "Tuberculosis Urogenitalis," alludes to the peritoneum as a great pathway of infection leading to the genital organs. His description of this method of infection applies to men, but if it holds good in that case, as it certainly seems to do, it carries double force in connection with the subject now under consideration: "My own observation leads me to the opinion that extension to the urogenital tract from the peritoneum is of much more frequent occurrence than is generally believed by surgeons. The fact that the peritoneal fluid is a medium in which bacilli may multiply; the dependent position of the rectovesical fold; the ease with which bacilli can pass through this thin membrane; its close relationship to the vesiculæ, the vasa deferentia, and the vesicoprostatic, venous, and lymphatic plexuses; and the well-known relationship between genito-urinary and peritoneal inflammation, — all afford explanation of the clinical observation. The insidious character of tubercular peritonitis, slowly

---

[1] *A System of Genito-urinary Diseases*, vol. i. p. 846.

and for a long time advancing, as it often does, without even a fever-rise to give warning of its presence, and with little or no pain, taken in connection with the fact that surgeons do not, as they might, take into consideration the fact that there is a peritoneal fold (vesicorectal) within reach of the fingertip, leads also to the belief that this course of tubercular progress is not so generally recognized as its importance justifies." The direct communication of the genital tract with the peritoneal cavity through the open extremity of the Fallopian tubes is an anatomical condition which would readily explain the great frequency with which tuberculosis of the peritoneum extends to the genital organs, and the constancy with which tubercular salpingitis is complicated by peritonitis of a similar type. It is well known that tubercular salpingitis begins usually at the distal end—another proof that tubercular peritonitis frequently stands in direct etiological relationship with genital tuberculosis.

From the opinions that have been quoted relative to the manner of infection in genital tuberculosis it becomes apparent that the genital organs are reached in one of the following ways:

1. In primary tuberculosis, by direct infection in which the external genitals may or may not be affected. In the former case the affection is prone to extend to the more distant genital organs by surface extension or through the lymphatic channels. In the latter case the tubercle bacilli deposited in the vagina or introduced into the uterine cavity are conveyed, like the spermatozoa, to the upper portion of the genital tract and even to the ovary itself.

In this event the essential microbic cause of tuberculosis will locate and produce its specific pathogenic effect wherever the local conditions present the necessary *locus minoris resistentiæ*. It is also reasonable to assume that tubercle bacilli from the peritoneal cavity may reach the Fallopian tubes in sufficient number and adequate virulence to produce tubercular peritonitis without having caused any antecedent demonstrable localized lesions on their way to the genital organs. Again, bacilli floating in the general circulation may locate in any part of the genital apparatus prepared for their action by congenital or acquired defects, without the presence of tubercular affections in any other organ or part of the body.

2. In secondary tuberculosis of the genital organs the affection occurs as a complication of antecedent tubercular processes in adjacent or distant organs, in the first instance the disease being caused by surface extension (peritoneum), by the rupture of a tubercular abscess into any portion of the genital tract, or, what is probably most frequently the case, the genital organs are reached through the general circulation or the lymphatic channels, when the genital disease appears as a single link of a long chain of pathological processes. We shall have occasion to call attention again to the different agencies of infection in considering separately the etiology of tuberculosis of the different organs of generation.

Frequency.—The greatest discrepancies of opinion exist among pathologists and operators relative to the frequency with which the female generative organs are affected by tubercular disease. This striking difference in the results of observations may be partly explained by the social position of the patients and by their place of residence, but it is largely due to imperfect methods of examination. In obscure cases a microscopical examination of the tissues and the demonstration of the tubercle bacillus in the inflammatory product must be made, and it is only by these diagnostic resources that a final and reliable differentiation can be made in many cases between a tubercular and a non-tubercular inflammatory lesion. According to Pollock, Schramm, and von Winckel, tuberculosis of the female genitals occurs in about 1 per cent. of all post-mortem examinations on women. Autopsies on phthisical women show the greatest discrepancy between the various observers, as will be seen from the following table, collected and arranged by Williams in his excellent monograph, already referred to:

Courty found 1 case in every 100 autopsies on phthisical women.
Louis " " " 66 " "
Cornil " " " 50–60 " "
Kiwisch " " " 40 " "
Mosler " " " 40 " "
Schramm " " " 24 " "
Nimias and Christoforis " 12 " "

That is to say, genital tuberculosis has been observed in from 1 to 8½ per cent. of all autopsies on phthisical women. The statements of operators as to its frequency also vary very much. Edebohls has met with 6 cases in 157 laparotomies for all causes, or 4 per cent.; Martin, 9 cases in 287 operations, or 3 per cent.; and in 137 laparotomies performed at Johns Hopkins Hospital genital tuberculosis was found in 2 cases at the time of operation, or 1½ per cent. In the writer's experience, including nearly 2000 laparotomies for various indications, tuberculosis of the female genital organs was noted 19 or 20 times. This percentage of tubercular affections would probably be considerably increased if the removed diseased organs had been submitted to a critical examination. Future experience and investigations will undoubtedly greatly increase the figures given heretofore in showing the relative frequency of tuberculosis of the female genital organs.

General Remarks on the Pathology of Tuberculosis of the Female Generative Organs.— The specific effect of the bacillus of tuberculosis on the tissues, irrespective of their structure and function, is to produce a chronic inflammation with certain retrograde tissue-metamorphoses which characterize the tubercular product. The histology of the primary tubercle in any part of the genital tract is almost identical with that of tuberculosis in any other organ. The giant cell and the epithelioid cells furnish here as elsewhere, the essential cellular elements of the nodule. Early avascularity of the inflammatory product characterizes here, as elsewhere, the tubercular process. The disease begins, as it does in any other organ, by the formation of gray miliary nodules, which may or may not undergo the typical retrograde metamorphoses—coagulation-necrosis, caseation, and liquefaction of the caseous material, an occurrence which precedes ulceration and abscess-formation. One of the principal features of the inflammatory product here, as compared with similar processes in most of the other organs, is that caseation is either entirely absent or sets in much more slowly than in most of the other organs. Caseation is an infrequent occurrence in any part of the mucous lining of the genital tract with the exception of the Fallopian tubes. It begins early and is a conspicuous pathological feature in ovarian tuberculosis. One of the striking features of tuberculosis of the peritoneum and the retroperitoneal lymphatic glands revealed by post-mortem examinations and operations is either an entire absence or incomplete caseation of the tubercular material. In intestinal tuberculosis the mesenteric glands undergo very early and speedy caseation, this condition forming a striking contrast to the state of the glands as found in the majority of cases in peritoneal and genital tuberculosis. Tubercular lesions of the mucous membrane lining the genital tract usually give rise to the formation of multiple ulcers, which increase more or less rapidly in size, become confluent, and thus give rise to large surface defects. As a rule, the ulcers are superficial, but in some cases which manifest a special aptitude for tubercular infection the ulcers increase very rapidly in size and penetrate the deep tissues. In tuberculosis of the external genitals ulceration is often preceded by great swelling—a condition described by former dermatologists as lupus vulgaris. In all tubercular surface lesions the destruction of the epithelial lining is followed by mixed infection with pus microbes—a complication which invariably hastens the destructive process and aggravates the general condition of the patient by the entrance into the circulation of septic material in greater or lesser quantities. With rare exceptions the tubercular process manifests an intrinsic tendency to local diffusion and regional and general dissemination. In the neighborhood of tubercular ulcers and abscesses there can always be found miliary tubercles, which in the course of time undergo degenerative changes, thus increasing step by step the area of infection. Spontaneous healing of tubercular affections of the female genital organs by cretefaction, cicatrization, or encapsulation, according to some authors, never takes place, and according to a few observers is of extremely rare occurrence.

Diagnosis.—The diagnosis of tuberculosis of any part of the genital apparatus is frequently uncertain, and often impossible, from the signs and symptoms presented during life or before operation. The coexistence of tuberculosis in other organs less difficult of detection is often of great diagnostic value. A careful study of the clinical history is of great importance, and should never be neglected. Lesions of the genital organs complicated by chronic peritonitis should receive careful attention with a view to ascertaining their probable tubercular nature. In doubtful affections of the external genital organs the use of the microscope is essential in the majority of cases for the purpose of settling the diagnosis definitely. If the lesion is tubercular, section of fragments of tissue reserved for diagnostic purposes will show under the microscope the typical structure of tubercle-tissue, and in some stained sections will reveal the presence of tubercle bacilli. In suspected tuberculosis of the endometrium search for tubercle bacilli in the uterine discharges should be made, and if this should prove negative, scrapings should be examined under the microscope to prove the tubercular or non-tubercular nature of the intrauterine disease. In tubercular salpingitis the use of the thermometer has proved of great diagnostic value in the experience of the writer. A constant slight rise in the evening temperature and a normal or slightly subnormal morning temperature are very suggestive of the tubercular nature of the tubal affection. In tubal disease without obliteration on the uterine side some of the secretions may escape into the uterus, and in the discharge from that organ tubercle bacilli may be found, so that the presence of tubercle bacilli in the uterine discharge does not necessarily prove that the endometrium is the seat of the tubercular affection. "To Babes (1883) belongs the credit of being the first to demonstrate tubercle bacilli in vaginal secretions, and since then the method has passed into routine practice. By this means Derville, Jouin, and others have been able to diagnose with certainty cases of tuberculosis of the uterus in which there was apparently no trace of tuberculosis elsewhere; and Derville states that Nocard has diagnosed tuberculosis of a cow's uterus by finding tubercle bacilli in the vaginal secretion, and that the diagnosis was verified at the autopsy" (Williams). Bimanual examination of the ovaries and tubes, even if the patient is placed under the influence of an anesthetic, cannot be relied upon in making a positive distinction between tubercular affections of these organs and other inflammatory diseases. The best such an examination can usually do is to prove the necessity of an operation, and leave the absolute diagnosis in the hands of the operator. In recapitulation, it may be said that the practitioner must base his probable diagnosis of genital tuberculosis on a thorough study of the clinical history of the case, a careful examination of all organs the frequent seat of tuberculosis, the use of the clinical thermometer, and microscopical examination of the vaginal discharge and of fragments of tissue, to which may be added inoculation-experiments by implanting or injecting into the peritoneal cavity or into the subcutaneous tissue of rabbits fragments of the suspected tissue or uterine discharge as a last and very valuable diagnostic resource.

Prognosis.—The prognosis of genital tuberculosis is always grave. If the disease is secondary to pulmonary or intestinal tuberculosis, life is cut short by the primary disease, not materially influenced by the genital complication. Unfortunately, the cases of primary tuberculosis of the genital organs seldom come under the observation of the surgeon before the disease has passed beyond the limits of a radical operation. From a prognostic standpoint, the most favorable cases are those in which limited tubal disease accompanied by a circumscribed hydrops of the peritoneal cavity is subjected to early abdominal section, with removal of the tubercular adnexa, iodoformization, and drainage. There can be no question that in many cases of tubercular peritonitis treated successfully by laparotomy and drainage the tubes were implicated in the tubercular process, and that the operative intervention exerted a favorable influence in arresting or retarding the tubal affection, rendering the contents of the tubes harmless by cretefaction or encapsulation. Spontaneous cures by similar processes are not impossible, but must be extremely rare.

Treatment.—The surgeon is too apt to forget that persons suffering from genital tuberculosis require the same careful general treatment as patients suffering from pulmonary tuberculosis. The strength of the patient must be maintained, and if possible increased, by a nutritious diet, the use of appropriate alcoholic stimulants, and change of climate. Such a course of treatment will increase the resistance to tubercular infection, and thus prove the most valuable means of arresting further local extension and preventing extension of the infection to distant, and perhaps more important, organs.

In anemic patients some preparation of iron, especially ferric iodide, will prove of great value. Among the antibacillary agents that have proved themselves most efficient in the experience of the writer, guaiacol and creosote deserve special mention. Both of these agents possess valuable therapeutic properties. Their use in gradually increasing doses should be continued for at least six months to a year. If digestion is unfavorably affected by long-continued use, the dosage should be diminished temporarily, to be followed again by a gradual increase. The writer is partial to guaiacol, but if this drug is not borne well by the stomach, some of its preparations—for instance, the carbonate—should be substituted for it. The best vehicle for the administration of guaiacol is milk. The drug should be given four times a day—before each meal and at bedtime—beginning with 5 drops and increasing to 10 drops or more. The guaiacol appears to neutralize or render harmless the toxins of the bacilli of tuberculosis, and by doing so exerts a curative effect upon the local lesions as well as upon the entire organism.

Change of climate is as salutary in the treatment of genital as in that of pulmonary tuberculosis. Absolute sexual rest must be enforced. Disregard of

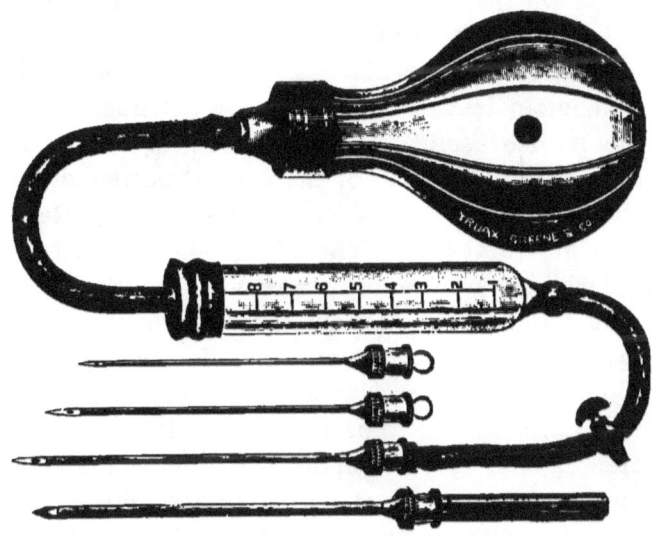

FIG. 9.—Senn's syringe for iodoform injections.

this advice is sure to be followed by speedy aggravation of the affection. Of the local remedies, iodoform and balsam of Peru are of special value. The former possesses valuable curative properties in the treatment of uncomplicated tuberculosis, and is especially serviceable in the treatment of tubercular abscesses. Under strict antiseptic precautions the accessible tubercular abscesses should be tapped and their contents washed out with a 5 per cent. boracic acid solution; after this from 2 to 4 drams of a 10 per cent. iodoform-glycerin emulsion should be injected (see Fig. 9), and upon withdrawal of the cannula the puncture should be sealed with iodoform collodion and a small pledget of aseptic cotton. These tappings should be repeated every week or two. If this treatment proves useful, three or four sittings will usually suffice. The first indication of the favorable action of the iodoform is furnished by the contents of the abscess, which become viscid and contain less of the tubercular detritus. Iodoform is also useful in the treatment of tubercular ulcers of the vulva and of the vaginal portion of the cervix uteri and in tubercular endometritis, after a thorough curettage. The sharp spoon must be used with sufficient energy to remove the tubercular product and to eliminate the cause of the secondary mixed infection. Iodoform is inert in the treatment of suppurating tubercular lesions. Iodoform is not destructive to the tubercle bacilli, but exerts a potent inhibitory effect upon them, and at the same time serves a useful purpose as an energetic tissue-stimulant by the action of which an active process of phagocytosis and repair is established. Balsam of Peru is indicated in the treatment of accessible tubercular ulcers in cases in which curettage is contraindicated. In such cases the occasional application of a 10 per cent. solution of chloride of zinc will prove of special value. The same solution may also be used as a parenchymatous injection around localized tubercular foci.

The operative treatment of genital tuberculosis will be considered in the description of the disease as it occurs in each one of the organs. The treatment of primary genital tuberculosis will yield more satisfactory results as soon as we are in a position to make an early and positive diagnosis.

PART IV.

TUBERCULOSIS OF THE VULVA.

ALMOST all authors agree that tuberculosis of the vulva is occasionally met with as a primary disease, although Hebra distinctly asserted that lupus of the vulva is never of primary occurrence, its lesions, when here present, having universally invaded the genital organs by extension from the thigh; and yet in 48 cases of lupus of the vulva collected by Peckham, a crural origin and subsequent extension to the vulva are not once recorded. In 1 case there is a record of simultaneous disease of the left thigh and the right labium; in 2 other cases there was ulceration of the groin, resulting from opening of abscesses in the inguinal glands; in as many more cases, as might be anticipated, there were isolated groups of pustules or nodules over the buttocks—the rare concurrent symptoms, in fact, of tertiary syphilis.

Taylor[1] mentions 3 cases in which the ulcers began just beyond the external genital organs and involved the vulva in their extension. They had finely and coarsely granular, papillomatous, and

[1] *New York Medical Journal*, Jan. 4, 1891.

even fungating surfaces, were encircled by hard, somewhat everted margins, and secreted an abundance of pus. In their initial stage they were round or oval tubercles, of a deep, even violaceous-red color, which soon broke down into ulceration. In 1 case the disease was associated with pulmonary tuberculosis.

The fact that tubercular infection of the genital organs takes place more frequently from within than from without becomes apparent from the singular immunity to this disease shown by those portions of the genital tract most exposed to outward infection. The rarity of tuberculosis of the vulva and the vagina as compared with those organs of generation more remote and better protected against external infection may, however, be due to a lesser degree of susceptibility to tubercular infection, so that the essential cause of infection may pass over the lower portions of the genital tract without producing disease, and reach the more important organs, where the tubercle bacilli come in contact with a soil better adapted to the exercise of their pathogenic effects.

While tuberculosis of the vulva is rare, and the reported authentic cases in literature are few, we have every reason to believe that the clinical material would be greatly increased if the cases described as lupus and *esthiomene* were classified, as they should be, with tubercular affections. To establish the diagnosis of tuberculosis of the vulva beyond the peradventure of doubt the presence of the tubercle bacillus or the histological structure of tubercle-tissue in the affected part must be demonstrated by the use of the microscope. Measured by such requirements, the cases of genuine vulvar tuberculosis recorded in literature are few. Many of the gynecologists have described tuberculosis of the vulva under the head of lupus. Duncan distinguishes two principal forms: 1. Lupus minimus; 2. Lupus maximus. Mann[1] divides lupus of the vulva into three classes: 1. Lupus serpiginosus, ulcerans, or exulcerans; 2. Lupus perforans; 3. Lupus prominens. Evidently many different things have been included in just such classifications. It is to be hoped that dermatologists, surgeons, and gynecologists will in the future exercise sufficient care in classifying ulcerative lesions of the vulva upon a bacteriological or histological basis, and as soon as this is done tubercular affections of the vulva will receive more attention in our current medical literature and text-books.

It is well to introduce this part of our subject by a brief description of the few cases of undoubted tuberculosis of the vulva recorded in literature.

Cayla[2] published a case of tubercular ulceration of the external genital organs in the form of deep ulcerations of the labia and of the entrance to the vagina. In the tissues of the ulcerated surfaces were found numerous minute nodules which under the microscope showed the typical structure of tubercle. The patient was afflicted at the same time with pulmonary tuberculosis, and it is very probable that infection occurred from the tubercular sputum.

---

[1] *American System of Gynecology*, vol. i. p. 520.
[2] *Progrès Méd.*, 1881, No. 33.

Deschamp's[1] case was that of a woman twenty-five years old suffering from advanced pulmonary phthisis, who, four months before she came under observation, injured her vulva by a fall down stairs. Soon after the injury the genital lesion made its appearance by a leucorrheal discharge and intense itching about the vulva. Examination revealed an extensive ulceration which involved the left lesser labium and the fourchette. The affected tissue under the microscope showed the typical structure of tubercle, and fragments of tissue inoculated into rabbits produced tuberculosis. The patient died three months later. Post-mortem examination showed that tuberculosis was limited to the lungs and the external genitals.

Montgomery[2] observed a case of tuberculosis of the vulva in a colored woman thirty years of age, the mother of four children. The first two children died at an early age from marasmus. She had one miscarriage five years ago. No hereditary disposition to tuberculosis existed. The disease commenced five years ago. At the time the examination was made both labia were swollen, the right more than the left. The mucous membrane within the vulva on the right side was excavated, presenting a furrow, a quarter of an inch deep and a third of an inch wide, extending from the posterior surface of the vagina forward to the anterior wall, and upward in the center the distance of three-fourths

---

[1] "Étude sur quelques ulcerations rare et non vénériennes de la vulve et du Vagin." *Archives de Toxicologie*, etc., 1885.
[2] "Tuberculosis of the Vulva." *International Clinics*, 1895, vol. iii, p. 280.

of an inch. The induration and edema around this ulcerated surface were quite extensive. Pain and tenderness were slight. No glandular involvement. No pulmonary tuberculosis. A number of the relatives of her husband had died of phthisis. Her husband was not very healthy, and complained very much of his back. The author made a diagnosis by exclusion, and expressed the belief that this might be a case of direct infection from husband to wife.

Chiari[1] reported a case in a woman thirty years old, the subject of tuberculosis of the lungs and extensive tubercular ulcers in the rectum. The ulcer of the vulva was large and had extended to the vagina. Post-mortem examination showed that the internal genital organs were intact. It was surmised that the vulvar disease was secondary to the rectal ulceration.

The often-quoted case of Zweigbaum[2] differs from the cases already described in that the disease commenced in the vagina, the extensive disease of the vulva being caused by extension of the vaginal ulceration. The patient died of pulmonary tuberculosis, and the post-mortem revealed that the uterus and the adnexa were intact. In nearly all the cases reported the lungs were the primary seat of the tubercular affection, and the pulmonary disease the direct cause of death. It would be reasonable to draw the inference, from the clinical history and the

---

[1] "Ueber den Befund ausgedehnter tuberculöser Ulceration in der Vulva und Vagina." *Vierteljahrsschrift f. Dermatologie und Syphilis*, 1886, p. 341.

[2] "Tuberculöse Geschwüre der Vulva, Scheide und der Vaginalportion." *Centralblatt f. Gynäkologie*, 1888, p. 494.

results of post-mortem examination, that the vulvar disease was caused in most of the cases by infection from without by tubercular sputum. In one case an injury of the vagina probably served as an infection-atrium.

Pathology.—Tuberculosis of the parts of the female genitals that have the structure of the external skin resembles in its clinical aspects and pathology the same disease as it occurs in the skin. Thin asserts that in tuberculosis of the vulva the process is very diffuse, like ordinary chronic inflammation, the greater number of cells being in the neighborhood of blood-vessels. Here we find a thick infiltration of leucocytes, epithelioid cells, and giant cells. Early and extensive ulceration is the rule in such cases. The ulcers may extend from the vulva to the vagina and from the vagina to the vulva. The tubercle bacilli may be found in the tissues after proper staining, but sometimes they are so scanty or so difficult of detection that a large number of sections (47 in one case—Cushing) may be necessary before they can be recognized.

The disease begins usually with a dull-red, livid discoloration, associated with marked infiltration and hypertrophy of the skin. There appear in some instances flattened light-yellow nodules, the size of a pigeon's egg, the cut surfaces of which present the appearance of granulation-tissue. On the surface of reddish-blue masses appear deep-red points the size of a lentil. As pain is usually absent or slight, the affection is often overlooked for a long time by the patient, until ulceration sets in. This is preceded by caseation and softening of the tubercular material. Multiple points of ulceration are frequently seen, which by coalescence form in a short time large and often deep surface defects. In the hypertrophic form of the disease ulceration takes place late, and extension of the disease is less rapid than in cases in which the infiltration-product undergoes early and speedy retrograde changes. Usually the tubercular ulcers in this locality are shallow; their margins are irregular, sharply cut, and slightly raised above the level of the surrounding healthy skin. The ulcers do not bleed readily, and when the yellowish-gray deposit is removed the edematous pale granulations are exposed. In the granulations and at some distance from the surface minute miliary tubercles can be found, which step by step invade the adjacent tissues and enlarge the field of infection. The margins and base of the ulcer lack the characteristic hardness so constantly present in carcinoma. Attempts at healing of the ulcerated surface are frequently seen, but the new scar-tissue again breaks down and the ulcer increases progressively in size. The discharge from the ulcer is a scanty clear serum, so that crust-formation is usually absent. Regional infection is generally absent, and the internal genital organs are usually found intact.

As patients suffering from vulvar tuberculosis are nearly always afflicted with pulmonary tuberculosis, death usually results from the latter affection, and before a sufficient time has elapsed for the extension of the disease to the internal organs of generation.

Diagnosis.—Tuberculosis of the vulva is most likely to be mistaken for syphilis or carcinoma. A careful examination of the internal organs should always be made for the purpose of detecting the primary seat of the tubercular process, should the disease exist in any other part of the body. From syphilis the affection is distinguished with some difficulty, but the history of the case and the effect of treatment will prove of diagnostic value. Phagedenic chancre is much more rapid in its growth, and the color of the base of the ulcer is more gray and yellowish-white than in a tubercular ulcer, and the granulations are of a brighter red color. Chancroid has none of the thickening of the surrounding tissues, and tubercles in the inflammatory product are absent. In syphilitic affections the inguinal glands are always enlarged, while this is seldom the case in tuberculosis. In doubtful cases a resort to the microscope becomes essential in making a positive differential diagnosis. The existence of multiple ulcers speaks in favor of tuberculosis and against carcinoma or syphilis.

Treatment.—In most cases of tuberculosis of the vulva the patients are the victims of pulmonary tuberculosis, and if the primary disease is far advanced, the local treatment should be purely palliative. In those rare cases in which the vulvar disease presents itself as an isolated and primary affection no time should be lost in placing the patient upon appropriate constitutional treatment and subjecting the lesion to efficient local treatment. Doutrelepont in 1884 made use of bichloride of mercury in the treatment of tuberculosis of the skin, and with some success. Others tried, with equally good results, a variety of chemical disinfectants for topical application, such as carbolic, sulphurous, and salicylic acids. Hutchinson strongly recommends sulphurous acid; Marshall, salicylic acid. Should such simple treatment prove unsatisfactory, it would be advisable in proper cases either to excise the ulcer or to remove the granulation-tissue with a sharp spoon, and after thorough disinfection to treat the wound with iodoform and pack with iodoform gauze. Healing of large granulating surface defects should be hastened by skin-grafting.

PART V.

TUBERCULOSIS OF THE VAGINA.

STATISTICS appear to prove that tubercular affections of the genital tract increase in frequency in proportion to the distance from the *introitus vaginæ*, the vulva being rarely affected, and the Fallopian tubes being the most frequent seat of the disease. Geil[1] found, in 45 cases of tuberculosis of the uterus, the vagina affected only in 3. He is of the opinion that tuberculosis occurs most frequently in the tubes, next in the uterus, and least frequently in the vagina. Vaginal tuberculosis occurs elsewhere as either a primary or a secondary affection in the form of miliary tuberculosis or tubercular ulceration. The primary form of the disease is probably nearly always the result of infection from without, the secondary variety constituting a continuation of a tubercular endometritis in a downward direction; or in rare instances it is caused by infection with tubercle bacilli from tubercular tubes, the intervening uterine mucous membrane remaining intact. In a few isolated cases it was caused by the perforation of a tubercular lesion of an adjacent organ into

---

[1] " Ueber die Tuberculose der weiblichen Geschlechtsorgane.". Dissertation, Erlangen, 1851.

the vagina. In miliary tuberculosis of the vaginal mucous membrane the minute nodules infiltrate the mucosa. The tubercles are of an opaque grayish color, except those which have undergone partial caseation; these present a yellowish color. With the caseation of the tubercle tissue the overlying epithelial lining is destroyed, leaving a ragged ulcerated surface.

Clinical Cases.—H. Thompson[1] reports the case of a girl fifteen years old, previously in good health, who died after an illness of five days' duration. The symptoms during life pointed to the lungs as the probable seat of the disease. The post-mortem revealed miliary tuberculosis of the lungs, diaphragm, liver, spleen, kidneys, and meninges. An imperforate hymen occluded the vagina completely. The latter was dilated into a sac 25 centimeters in circumference, and contained 750–790 grams of a grumous, fetid, semi-fluid mass evidently composed largely of retained menses. The vaginal portion of the uterus and the os uteri were covered with tubercular granulations; the vaginal mucous membrane was extremely vascular and was infiltrated with minute tubercles.

Breisky[2] saw a case of tuberculosis of the vagina in a woman, advanced in years, who had been operated upon for ovarian cyst, which upon examination also showed tubercular lesions.

Zweigbaum[3] (cited before) mentions a case of

---

[1] "Case of Acute Tubercular Disease with Occlusion of the Vagina." *London Lancet*, 1872, No. 5.

[2] *Handbuch der Frauenkrankheiten*, Billroth u. Luecke, B. iii. p. 72.

[3] *Virchow's Jahresbericht*, 1887, B. ii. p. 713.

tuberculosis of the vulva, vagina, and vaginal portion of the cervix uteri. The patient was thirty-two years of age. The posterior vaginal wall was the seat of a cavernous ulcer with elevated, hard, uneven margins; the surface of the ulcer was covered with a thick grayish secretion. Upon the vaginal portion of the cervix was a similar ulcer. The uterus was enlarged and tender upon pressure; the inguinal glands were swollen and hard but painless. The left pulmonary apex was infiltrated. Suppression of menses had existed for eight months. Later the vaginal process extended to the left labium minor, which in a short time was completely destroyed. Microscopic examination of tissue-fragments from the ulcers showed tubercle bacilli. Local treatment had little effect on the course of the disease. Death occurred five months later. The vaginal ulcer perforated the rectovaginal wall, causing a fistula. The author was not certain whether the vaginal disease was primary or secondary.

Klob[1] describes a case of multiple tubercular ulcers which he discovered in the cadaver of a woman sixty-seven years of age. The ulcers were of the size of a lentil, with undermined margins, the base being covered with a yellowish membrane. The tissues between the ulcers were freely infiltrated with gray and yellow tubercles. Of the remaining organs, the lungs, liver, and intestines were also the seat of tuberculosis.

Tuberculosis of the vagina has been met with

[1] *Op cit.*, p. 432.

and described by a number of other authors, but the cases cited are sufficient to illustrate some of the clinical aspects and the most salient points in the etiology of the affection.

Pathology.—The form of vaginal tuberculosis will depend largely on the manner of infection and the presence or absence of serious complications. In primary tuberculosis the disease pursues a very chronic course, the miliary nodules undergo caseation, and ulceration follows the softening and disintegration of the tubercular product. If the disease does not advance to this stage, it is always secondary to some other tubercular affection. The cervix uteri and the vaginal roof are the favorite locations for the infection. As in many other tubercular affections, in caseous vaginitis ulcers form at different points and by confluence form large surface defects. Not infrequently the small ulcers cast off the tubercular deposit, in which case confluence does not take place. The ulcers may be superficial or deep, with abrupt, steep margins, and the floor covered with a cheesy material under which is found a pavement of pale edematous granulations. Miliary tubercles can almost always be found near the margins and underneath the surface of the ulcer. If the ulcers penetrate deeply, the bladder and the rectum may be perforated, leading to a tubercular vesicovaginal or rectovaginal fistula, with the probability of the invaded organ becoming infected from the vagina

Etiology.—Until recently it was the general belief among pathologists that tuberculosis of the vagina is always secondary to tubercular endometritis.

Klebs[1] regards vaginal tuberculosis as being an affection always accompanied by far-advanced uterine tuberculosis. He also maintains that this disease is never met with in children, while the tubes are affected in children as well as in persons advanced in years. These views are no longer tenable, as it has been ascertained that the vagina may become infected from different sources, and vaginal tuberculosis has been found in children of tender age. The structure of the normal vaginal mucous membrane is such that this organ is anatomically not predisposed to tubercular infection. The mucous membrane of the vagina is a derivative of the epiblast, and the different layers of pavement-epithelium, when in a normal condition, furnish a good protection against direct infection. The resistance of the normal vagina to direct infection has been demonstrated by experiments on the lower animals. "In 1889, Cornil and Dobrolowsky stated that they had been able to produce tubercular endometritis by injecting pure cultures of tubercle bacilli into the vagina of rabbits, but thus far no one has substantiated their results. Indeed, later work tends to show that their results were erroneous, for in 1890 Oncarani stated that the introduction of tubercular material into the vagina of rabbits, even if previously irritated by rubbing with tincture of iodin, never produced genital tuberculosis" (Williams). Williams repeated these experiments, but in none of his cases was there any reaction on the part of the animals, and on killing them at periods of four

---

[1] *Handbuch der Pathologischen Anatomie*, B. i. p. 959.

to six weeks afterward he was unable to find the slightest trace of tuberculosis anywhere in the body. Cases of direct infection of the vagina by the introduction of tubercle bacilli by coitus or otherwise (tubercular sputum) are no longer lacking, so that this method of direct infection must be recognized in rare cases as established. Infection by coitus, as first suggested by Cohnheim (1879) and Verneuil (1883), can no longer be considered in the light of an imagination. Such a source of infection is possible in cases of tuberculosis of the external genitals of the male, and women have been known to contract the disease from phthisical husbands without any external tubercular affections. Jani[1] has shown the presence of tubercle bacilli in the testes and prostate glands of phthisical men without any evidence of disease of those organs. Thus in 8 cases of phthisis he found bacilli 5 times in the testicle and 4 times in the prostate, while none could be detected in the semen. It is not difficult to conceive that tubercle bacilli in pathogenic quantity could reach the vagina under such circumstances by coitus, and in the event of their being brought into contact with an appropriate soil, vaginal tuberculosis could originate in that way.

Abrasions and catarrhal and gonorrheal inflammation are no doubt occasionally responsible for the localization and disease-producing action of tubercle

---

[1] "Ueber das Vorkommen von Tuberkelbacillen im gesunden Genitalapparat bei Lungenschwindsucht, mit Bemerkungen über das Verhalten des Fötus bei acuter Allgemeiner Miliartuberculose der Mutter." *Virchow's Archiv*, B. ciii. p. 522.

bacilli in the vagina as well as in other parts of the genital tract. Vaginal infection produced in this manner may lead to tuberculosis of the internal genital organs by the tubercular infection following the lymphatics without causing a demonstrable lesion in the vagina. The most conclusive proofs of the transmission of tuberculosis by coitus have been brought out by Derville,[1] who in 8 cases of genital tuberculosis in women found bacilli in the secretions from 5, and in examining the husbands or lovers of these women found that all of them were afflicted with tubercular epididymitis. The introduction into the vagina of tubercle bacilli from other sources and such as are excreted by the patient and the extension of a tubercular affection from the vulva to the vagina, furnish other instances of infection from without.

Tuberculosis of the vagina is not always dependent upon direct infection from without, or, as has been claimed by Klebs, upon advanced tuberculosis of the uterus. In the cases reported by Virchow, Weigert, Klebs, and Kasewarowa the uterus was not affected. In such cases tuberculosis of some of the adjacent parts or of one or more of the abdominal organs is never absent; the Fallopian tubes, the urinary organs, the intestines, and the peritoneum show more or less advanced tuberculosis. In Virchow's case[2] the disease, which had its primary origin in the urinary tract, resulted in perforation of the vagina, which became infected through the fistula by the

---

[1] "De l'infection tuberculeuse par voie génitale chez la femme." *Thèse de Paris*, 1887.
[2] *Virchow's Archiv*, B. v. 1853.

irrigation of the vaginal mucous membrane by the tubercular urine.

Weigert[1] found in the cadaver of a woman sixty-seven years old, besides pulmonary tuberculosis of a moderate degree and tubercular ulcers of the ileum exquisite peritoneal tuberculosis. While the mucous membrane of the uterus was perfectly intact, the mucous membrane of the upper portion of the vagina was the seat of a tubercular process in all stages from miliary tuberculosis to the formation of cheesy material and ulceration. The nature of the disease was established by microscopic examinations.

Infection from tubercular tubes without implication of the uterus has already been referred to as a possible cause of vaginal tuberculosis.

Oppenheim,[2] from the observations of Orth, describes 7 cases of tuberculosis of the vagina, in 3 of which a normal uterus intervened between the tubercular tubes and the vaginal lesion. Weigert mentions a case in which the infection evidently extended from the peritoneum to the vagina. The peritoneum was the seat of advanced tuberculosis, while the lungs and the intestines were only slightly affected. In the vagina miliary tubercles and ulcers were found. Tuberculosis of the intestines with perforation may reach the vagina and cause secondary tuberculosis. Babes in 1883 described a case in which tubercular ulcers of the rectum perforated the

---

[1] "Tuberculosis Vaginæ." *Virchow's Archiv*, B. lxvii. 264.
[2] "Zur Kenntniss der Urogenitaltuberculose." Dissertation, Göttingen, 1889.

rectovaginal wall and caused secondary tuberculosis of the vagina.

Among the predisposing causes must be mentioned age, injuries, and antecedent inflammatory affections of the vaginal mucous membrane. The last two etiological factors furnish the necessary infection-atrium, through which the tubercle bacilli find their way into the vascular portion of the vaginal wall. While the age of active sexual life is most apt to tubercular infection of the vagina, owing to greater liability to injuries and inflammatory affections of the mucous membrane, it appears that no age is entirely exempt. Parrot describes a case of vaginal tuberculosis in a girl seven years old, and Kiwisch found the same disease in a woman seventy-nine years of age. Demme[1] found in children one to two years of age vulvar and vaginal tuberculosis. Children thus affected always died sooner or later of some other intercurrent tubercular disease.

Diagnosis.—As the lesion in tuberculosis of the vagina is readily accessible to inspection and palpation, the diagnostic difficulties encountered in recognizing this disease are much less than in tubercular affections of the internal genital organs. The remarks made on diagnosis of tuberculosis of the vulva apply with equal force to this part of the genital tract. The search for the primary source of the infection, the chronicity of the disease, and the characteristic appearance of the miliary infiltration or tubercular ulceration will usually enable the careful practitioner to make a correct diagnosis at

---

[1] *Bericht über die Thätigkeit des Jenner'schen Kinderspitals,* Bern, 1887.

an early stage of the disease. One of the conspicuous clinical features of the disease is the almost total lack of pain and tenderness, which will distinguish it from acute and other forms of chronic inflammation. Granular vaginitis, which might be mistaken for miliary tuberculosis, is very frequently associated with gonorrhea and pregnancy, while miliary tuberculosis, when diffuse, always occurs in phthisical women. Carcinoma of the vagina as a primary disease is very rare, and always presents the characteristic induration of the base and margins of the ulcer, which readily distinguishes it from the ulcerating form of vaginal tuberculosis. Primary and ulcerative syphilides are differentiated from tubercular affections by a study of the clinical history, the presence of general hyperplasia of the lymphatic glands, and in doubtful cases by the administration for a few weeks of antisyphilitic remedies. As a final means of diagnosis, fragments of tissue should be subjected to microscopic examination.

Treatment.—In miliary diffuse tuberculosis of the vagina the treatment must necessarily be limited to palliative measures. In primary tuberculosis the usual general and local measures employed in the treatment of accessible tubercular affections are indicated. Daurios[1] recommends for the smaller ulcerations repeated applications of the tincture of iodin, iodoform, and chromic acid. This treatment will not suffice for deeper ulcerations, which require the intervention of the surgeon. If the ulcers are located on the cervix, amputation of that portion of

---

[1] *Revue médico-chirurgicale des maladies des femmes*, March 25, 1891.

the uterus, or removal of the entire organ by the vaginal route, may become necessary. In progressive deep ulcers, incision, deep cauterization with the Paquelin cautery, and curettage followed by iodoformization and packing with iodoform gauze are the measures that should be relied upon. Appropriate general treatment should never be neglected. In secondary tuberculosis of the vagina following perforation of a tubercular abscess or ulcer of an adjacent part or organ, the primary affection must receive proper attention.

PART VI.

TUBERCULOSIS OF THE UTERUS.

The uterus comes next to the Fallopian tubes in the frequency with which it is the seat of the disease in genital tuberculosis. This organ has been found quite frequently affected at necropsies on phthisical women, but it is only recently that gynecologists have had an opportunity to study the pathological conditions of specimens removed during life.

T. S. Cullen, in a paper on "Tuberculosis of the Endometrium," recently published,[1] describes three specimens of tubercular endometritis removed by Kelly during eighteen months. These cases go to show that uterine tuberculosis is much more frequent than has generally been supposed and as would appear from post-mortem records. During the early stages of the disease the textural changes are so slight that no correct diagnosis is possible without the use of the microscope. Rokitansky and Kiwisch found among 40 women who had died of tuberculosis only 1 case of tuberculosis of the uterus; according to Winckel, tuberculosis of the tubes and uterus is found in 1 per cent. of all post-mortems on

[1] *Johns Hopkins Hospital Reports*, vol. iv. Nos. 7 and 8.

women. Careful examination of the specimens removed by operative intervention will prove the percentage of uterine tuberculosis to be much greater.

Primary tuberculosis of the uterus is very rare as compared with the frequency with which this organ is involved secondarily. The body of the uterus is much more frequently the seat of the tubercular process than the cervix. Tuberculosis of the vaginal portion of the uterus is so rare that Rokitansky and Lebert denied its existence. There are, however, a number of well-authenticated cases on record in which the disease was limited to this part of the uterus. In 1885, Späth,[1] in searching the literature, found 6 in 119 cases of genital tuberculosis. J. D. Williams[2] reports 2 cases of undoubted tubercular disease of the vaginal portion of the cervix uteri. In both cases the cervix was the seat of ulcers which under microscopic examination revealed tubercle nodules and bacilli. The disease was limited to the vaginal portion of the cervix and the adjoining vaginal mucous membrane, with no involvement of the rest of the genital tract. Both cases were complicated by extensive pulmonary disease.

Emanuel[3] records a case of tuberculosis of the cervix that extended to the vagina but did not affect the uterus or the tubes. The uterus was removed by abdominal section, but the patient died of shock

---

[1] "Ueber die Tuberculose der weiblichen Genitalien." Dissertation, Strassburg, 1885.

[2] *Medical Press and Circular*, Sept. 5, 1894.

[3] "Beitrag zur Lehre von der Uterustuberculose." *Zeitschrift f. Geburtshülfe u. Gynäkologie*, B. 29, p. 135.

soon after the operation. The lungs were free from tubercle, but the spleen, liver, kidneys, peritoneum, and mesenteric glands were involved. The cervical lesion resembled very closely carcinoma, so that the differential diagnosis had to be made by the aid of the microscope. The author is inclined to the belief that in this case infection occurred by coitus.

E. Fränkel[1] made a post-mortem on a woman twenty-eight years old who died of tubercular spondylitis and its result, compression of the cord. Examination of the genital organs revealed very slight swelling of the Fallopian tubes, but complete destruction of the mucous membrane by tubercular salpingitis. The body of the uterus was normal. The cervical canal from the internal os was filled with papillary growths 1½ centimetres in length; they also involved a portion of the posterior lip of the portio vaginalis Upon the exposed part of the cervix the tubercular product presented itself in the form of granulations instead of papillary growths. In the cervical canal the disease resembled very closely cauliflower carcinoma, as in the case reported in 1889 by Cornil, and operated upon by Péan for carcinoma. Giant cells and tubercle bacilli were found in the papillary excrescences.

Fränkel recognizes three forms of tuberculosis of the cervix uteri: 1. Miliary tubercles; 2. Diffuse caseous infiltrations; 3. Papillary growths. The duration of the secondary tubercular cervicitis of his case is unknown. It is difficult to explain in what

---

[1] "Beitrag zur Lehre von der Uterustuberculose papilläre, Cervictuberculose." *Centralbl. f. Gynäkologie*, 1896, No. 39.

manner the disease involved tubes and cervix without implicating the uterus.

R. Emanuel[1] reports a case of tuberculosis of the cervix uteri in which the swelling attained the size of an apple, and which in its microscopic appearance and clinical course led to the diagnosis of malignant disease—either carcinoma or sarcoma.

The endometrium is much more frequently affected by tuberculosis than the vaginal portion of the uterus, for anatomical reasons, and because of the greater frequency with which lesions occur that determine localization of tubercle bacilli, such as pregnancy, the puerperium, menstruation, and catarrhal inflammation. J. W. Williams describes three pathological varieties: 1. Miliary tuberculosis, with or without ulcerations; 2. Chronic diffuse tuberculosis (caseous endometritis); 3. Chronic fibroid tuberculosis. Miliary tuberculosis of the uterine mucous membrane occurs in connection with grave tubercular affections of some other organ, and has on that account little clinical interest except to the obstetrician. as occasionally this disease develops soon after childbirth, is limited to the uterus, and might be mistaken for puerperal sepsis.

Rokitansky[2] describes a case of acute tuberculosis of the puerperal uterus. The uterus was large, and its walls were very thick; the mucous membrane was edematous and was infiltrated diffusely with gray and yellow nodules. The same kind of tubercular granulations invaded also the entire thickness

---

[1] *Zeitschrift f. Geburtshülfe und Gynäkologie*, B. xxix.
[2] *Allgemeine Wiener med. Zeitung*, 1860, No. 21.

of the muscular wall. Fatty degeneration of the muscular fibers was far advanced. The mucous membrane of both tubes was the seat of gray miliary tubercles, evidently the result of extension of the disease from the uterus to the tubes. Pulmonary tuberculosis and fatty liver were the complications found in other organs. The patient was thirty-four years old, and had given birth to an eight-months' child nineteen days before death. According to Rokitansky, tuberculization of the uterus occurred after delivery, in consequence of the loss of resistance to tubercular infection sustained by the uterus during the puerperal state.

In caseous endometritis the tubercular product undergoes the characteristic degenerative changes. The disease is chronic, and in the same specimen may be found miliary tubercles, more or less caseation, and ulceration. Fibroid tuberculosis is characterized here, as elsewhere, by the copious production of connective tissue, which retards the progress of the disease and imparts to it its most noted clinical feature—chronicity.

Pathology.—For practical purposes it would be sufficient to divide uterine tuberculosis of the uterus into two classes: (1) acute and (2) chronic. In diffuse miliary tuberculosis surgical treatment is out of the question, as the disease is invariably complicated by tubercular affections of other organs that contraindicate radical local treatment. In the chronic variety, whether it assumes a caseous or a fibroid type, the question of partial or complete hysterectomy may arise, and must be decided according to the merits of each individual case. From an etiological and pathological standpoint, however, the classification made by Williams should be generally adopted.

As has been stated before, isolated tuberculosis of the cervix uteri is a very rare pathological condition. Pozzi[1] was able to find the report of only 2 cases— one by Laboulène and the other by Cornil. The case described by Cornil is given in detail, and the illustrations accompanying the description of the case show in an excellent manner the histological appearance of the tubercular process in this part of the genital tract. "The tubercle nodules are at first covered by the normal layers of stratified pavement-epithelium, and consist of the characteristic histological elements—giant cells, epithelioid cells, and granulation-cells. Below the mucous membrane we find a small number of tubercular follicles situated in the midst of crossed muscular bundles. These muscular bundles are at a given point separated by embryonal connective tissue, which forms an island in the midst of which are one or several giant cells surrounded by epithelioid cells." If the mucous membrane of the cervical canal is affected, the glands are first enlarged and the connective tissue spaces are filled with small round cells. Under high power the connective tissue of the mucous membrane between the glands is seen to be infiltrated with small cells and characteristic giant cells. In this superficial form of tuberculosis of the cervix uteri no epithelioid cells could be discovered by Cornil. The minute tissue-changes observed in

[1] *Medical and Surgical Gynecology*, vol. ii. p. 198.

miliary tuberculosis precede caseation and fibroid changes which characterize the chronic form of uterine tuberculosis. In the endometrium, as elsewhere in the genital tract, the first evidences of the tubercular inflammation appear in the vascular portion of

FIG. 10.—Section from the fundus of a tubercular uterus; × 250. Three uterine glands can be seen. Near the center are three tubercles with a giant cell in the center of each. Around the giant cells are epithelioid cells and small round cells (T. S. Cullen).

the mucous membrane, in close proximity to bloodvessels and between the uterine glands. Here tubercles as described by Langhans are formed, varying in size from a millet-seed to that of a hempseed (see Fig. 10). The center of the nodule is usually formed by a giant cell with numerous peripheral nuclei, surrounded by a zone of epithelioid cells around which the connective-tissue perivascular spaces are infiltrated by small round cells. The giant cells are not always typical. "Some of the tubercles especially those at the top of the fundus, contain giant cells which are irregularly oval or round in contour, and which contain oval vesicular nuclei, arranged either around the margin of the cell, at one or both poles, or scattered promiscuously throughout the protoplasm" (Cullen). As the interstitial disease advances, the epithelial cells of the uterine glands take an active part in the tissue-proliferation. Cullen has shown that the gland-cells later take an active part in the production of tubercle-tissue, some of them, at least, being converted into epithelioid cells. The mural intraglandular tubercle projects at first in the glandular space, which in the course of time becomes packed with tubercular tissue, resulting in complete destruction of the gland. The tubercle nodules by pressure and undoubtedly by the local action of the toxins upon the adjacent capillary vessels prevent vascularization of the new product and thereby hasten degenerative changes. As the nodules reach the surface, the epithelial cells lining the mucous membrane are exposed to similar harmful influences. In some places the epithelial cells proliferate, forming small teat-like projections or delicate gland-like spaces before the caseous material has reached the surface. With the occurrence of caseation and softening of the tubercular product the overlying epithelial cells are destroyed and larger or smaller surface defects are produced. The mucous membrane covering the infected area is during the first stage of the disease abnormally vascular and somewhat thickened. These conditions are intensified if the affected organ before the tubercular affection has been the seat of catarrhal endometritis. In more advanced cases the size of the uterus is increased, its walls are thickened (and in the caseous form edematous), and the uterine cavity is enlarged. Hyperplasia of the uterine glands is a marked feature during the early stages of the disease. By confluence of the tubercles the areas of caseation and subsequent ulceration are increased.

As uterine tuberculosis is generally secondary to a similar affection of the tubes, the fundus of the organ is usually the primary location of the disease. From the fundus the process extends by continuity of surface downward in the direction of the internal os, and sometimes involves also the cervical mucous membrane. The primary tubercular ulcer of the endometrium is small, round or oval, with whitish-yellow infiltrated margins, the floor uneven and covered with cheesy material. By confluence of nodules and ulcers large surface defects are created.

Section through a primary tubercular ulcer exhibits all the essential histological structures of tubercle. On the surface may be seen the necrotic material, in which cell-remains and polynuclear leucocytes are visible. In the tissues underneath, giant, epithelioid, and small round cells mark the area where the disease is advancing deeper into the muscular wall of the uterus. Tubercle bacilli are very numerous in the caseous areas and in the giant cells; their presence can also be demonstrated in the adjacent glands and among the epithelioid cells. The advance of the disease in the direction of the muscular tissue is retarded in many cases by hyperplasia of the connective tissue, in which, however, the tubercular infiltration finally disseminates the process, so that ultimately the entire thickness of the muscular wall may become involved, leading in rare instances to the formation of large excavations or cavities. H. Cooper reports a case in which extensive tuberculosis of the uterine wall led to rupture. In some cases the tubercular process, when

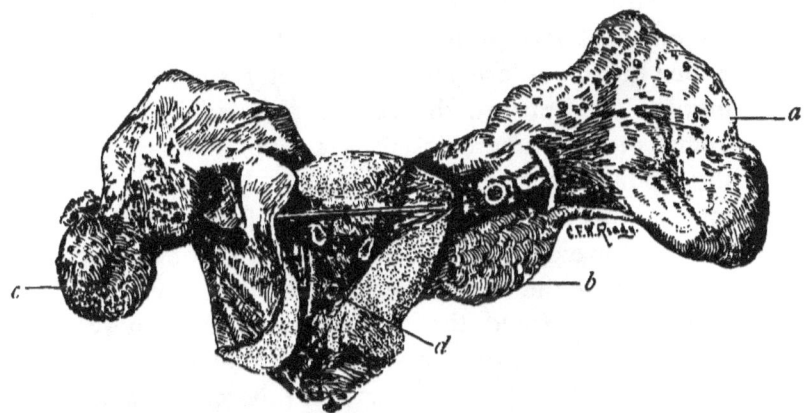

FIG. 11.—Tuberculosis of the uterus and the Fallopian tubes: a, right tube; b, ovary; c, left tube, left ovary imbedded in adhesion; d, tubercular ulcers of the endometrium, two at the fundus and one near the internal os (post-mortem specimen from a patient who died of pulmonary tuberculosis).

it reaches the cervical wall, causes inflammatory obstruction at the internal os, when retention of the tubercular material in the uterine cavity results in distention with a corresponding increase in the size of the organ. Cervical obstruction from this cause is more likely to occur in aged women; but Silcock met with a case in a child five years old. Krzywicki mentions two cases in women aged respectively seventy-one and eighty-three years. Quite frequently the uterine disease is followed by extension of the infection to the Fallopian tubes, or the tubercular endometritis follows tubal tuberculosis (see Fig. 11), or both the uterus and the Fallopian tubes are affected simultaneously. It is not always easy to determine, by examination of specimens, the chronological order of the diseases of these two organs when both are infected. The organ in which most caseation and ulceration are found is usually regarded as the one that was primarily involved. According to Klob,[1] the tubes are more frequently affected primarily. Contrary to the course pursued by carcinoma, uterine tuberculosis, as a rule, travels from the fundus in a downward direction. Very often the internal os constitutes the limit to its further extension. However, as in carcinoma, exceptions occur here also, as the disease may attack the cervix primarily, and extend thence to the uterine cavity; but such cases are quite rare. Klob examined a case in which he found in the cervix a tubercular cavity the size of a cherry, and he believes that the disease had its starting-point in one of Naboth's glands, from the walls of which the tubercular infection extended to the adjacent tissue.

In fibroid tuberculosis of the uterus caseation is limited, and the rapid extension of the disease is prevented by connective-tissue hyperplasia within and

---

[1] *Pathologische Anatomie der weiblichen Sexualorgane*, Vienna, 1864, p. 193.

around the tubercular area. Although this form has as yet not been described as a distinct variety of uterine tuberculosis, we must naturally infer that it exists, as the conditions in the uterus are similar to those found in the tubes, where the disease has been well described clinically and pathologically by Williams and Penrose.

Etiology.—Tubercular infection of the uterus may occur as a primary disease, by infection through the blood, or as a secondary affection, from the extension of a tubercular inflammation from the tubes or vagina; or, lastly, in the miliary form, as one of the results of miliary tuberculosis caused by tuberculosis of the lungs or any other organ anatomically disconnected with the uterus. As a primary localized lesion it is caused by infection from without, the tubercle bacilli being conveyed to the seat of the disease without causing any lesions of the lower portion of the genital tract. The vaginal portion of the uterus is less susceptible to tubercular infection than the endometrium; hence this portion of the uterus is more frequently the seat of primary tubercular disease. Williams believes that infection of the endometrium from below does not take place frequently; that it is due to the exfoliation of the superficial layers of the epithelial cells during menstruation; and that when it does occur, the infection has extended to the Fallopian tubes, where the tubercle bacilli can multiply unmolested, or the case is one in which the bacilli and the beginning tubercles were not carried away at the menstrual period. This explanation would, however, not hold good in cases of tuberculosis of the uterus such as those described in patients before and after the menstrual age. P. Maas[1] found in literature only 7 cases of tuberculosis of the genital organs in children. He gives an accurate account of a case in which the disease had its primary starting-point in the genital organs, and resulted in death by extension thence to other organs. He believes that genital tuberculosis, as a rule, occurs in consequence of coitus with tubercular individuals or because of the introduction of tubercle bacilli by fingers or instruments in making vaginal examinations, and attributes the rarity of the disease in children to the presence of an intact hymen. In the case fully described by Maas, infection appears to have taken place through the umbilicus, and the oldest tubercles were found in the connective tissue and serosa from the umbilicus in the direction of the peritoneal cavity. During the further development of the disease the infection probably extended from the peritoneum to the Fallopian tubes and uterus. The original infection-atrium was in all probability the open granulating umbilicus. Infection from the peritoneal cavity is possible without implicating the intervening tubes in cases in which the mucous membrane of the uterus is the seat of special localizing lesions, such as catarrhal endometritis or the puerperal state.

Extension of a tubercular affection of organs other than the tubes and vagina must be an extremely uncommon cause of tubercular endometritis.

[1] "Ueber die Tuberculose der weiblichen Genitalien im Kindesalter." Dissertation, Bonn, 1896.

Kaufmann[1] reports such a rare instance. A tubercular ulcer of the small intestine, after the formation of adhesions, perforated into a myomatous uterus, and through the fistulous communication infection of the uterus occurred. Another rare source of infection alluded to by Klob is tuberculosis of the retroperitoneal glands. In cases of simultaneous disease of the uterus and tubes the latter are nearly always more affected by the process than the uterus; according to Förster, Klob, Wernich, and others, it is probable that the tubes were the primary seat of the disease.

Remarkable is the frequency of the occurrence of uterine tuberculosis after the puerperium, at a time when the involution of the organ is nearly completed. In such cases the infection commences where the placenta was located, and at a time when the muscular fibers are undergoing fatty degeneration. That a tubercular uterus can conceive is shown by Cooper's case, in which a woman pregnant three months died from spontaneous rupture of the tubercular uterus. Grisolle made the statement that pregnancy exerts an inhibitory effect on the tubercular process, while the puerperium predisposes to infection. This opinion is not supported by Lebert, who bases his views on 33 cases, of which number only 13 were between the ages of twenty-five and thirty. The greater receptivity of the puerperal uterus for tubercular infection is now generally recognized. If tuberculosis has existed

---

[1] "Perforation eines tuberculösen Dünndarmgeschwürs in die durch Myoma deformirte Uterushöhle." *Arch. f. Gynäkologie*, B. xxix. p. 407.

before marriage, it is liable to recur most frequently during the second or third pregnancy. Hoffmann[1] observed in a young woman, soon after childbed, a chronic metritis. Tuberculosis was not hereditary in the family. During the course of the disease occurred an inflammation of the left sacro-iliac joint and the adjacent pelvic connective tissue, followed by caries and abscess-formation. About the same time pulmonary tuberculosis appeared, to which the patient succumbed. At the post-mortem there was found, besides pulmonary, joint-tuberculosis, and peritoneal tuberculosis, an advanced tuberculosis of the mucous membrane of the uterus and tubes, of about the same duration as the remaining tubercular affections. Breus[2] made a necropsy on a woman thirty-four years old, who eight months before had aborted during the fifth month of her second pregnancy, and had died of tubercular meningitis. In the right tube was found a cheesy focus; the mucous membrane of both tubes was studded thickly with tubercles. The uterus was very large. The endometrium was edematous and diffusely infiltrated with tubercles. The muscular wall was of a pale yellowish-red color and was infiltrated with miliary tubercles throughout. A few nodules were also found in the cervix and the parametrium. The ovaries and vagina were intact. Heidenthaler[3] reports a case of tubercular metritis of unusual inter-

---

[1] "Ein Fall von Tuberculose der Gebärmutter." Bayer. ärztl. Intelligenzblatt, 1867, No. 3.

[2] "Ueber Akute Tuberkulose des Uterus." Wiener med. Wochenschrift, 1877, No. 44.

[3] Wiener klinische Wochenschrift, August 21, 1891.

est, since in the majority of cases under similar circumstances the tubercles were of the miliary variety. The patient was twenty-nine years old, and had given birth to three children, one of whom had died of intestinal tuberculosis. The father of the patient died of phthisis. Two months after a miscarriage she began to have a bloody vaginal discharge, which later became purulent and fetid. With progressive emaciation abdominal pains with cramps in the lower limbs set in. Examination revealed pulmonary phthisis, and the left Fallopian tube was enlarged to the size of the little finger. On the anterior lip of the cervix there was a superficial ulcer as large as a quarter-dollar. A diagnosis was made of tuberculosis of the lungs, tube, corpus and cervix uteri, with probable tubercular enteritis. At the autopsy this diagnosis was confirmed in every respect. The entire endometrium was studded with cheesy nodules. The author believed that the tubes were the primary seat of the tubercular process, and that the wide dissemination was caused after the miscarriage.

The frequency with which infection of the uterus occurs after childbed is undoubtedly in part due to the increased susceptibility of the uterus when in that condition, and the increased risks of infection of the genital organs during delivery by the hands or instruments of the accoucheur, by dressings, and by the clothes of the patient—in fact, by anything contaminated with tubercle bacilli brought in contact with the genitals. There is, however, another and probably much more fruitful cause to explain tuberculosis of the puerperal uterus, and that is the localization of tubercle bacilli in the placenta during pregnancy. The wonderful changes that the circulation of the uterus undergoes during pregnancy, and the conditions caused by detachment of the placenta on the surface of the uterine wall, are well calculated to cause the localization of tubercle bacilli in pathogenic quantities in a soil favorable for their growth and multiplication. The involution of the blood-vessels and the fatty degeneration of the muscular fibers that take place soon after delivery cannot fail in reducing the resistance of the puerperal uterus to infection of all kinds. The direct transmission of tubercle bacilli from mother to child has been established by experiments and clinical observations. At the French Congress for the Study of Tuberculosis in 1891 Vignal gave the result of a large number of experiments made for the purpose of studying the direct transmission of tuberculosis from mother to fetus. Pregnant guinea-pigs were infected. In one series of experiments fragments of the fetus made into an emulsion with sterilized physiological solution of salt were used to infect other guinea-pigs, but the results were always negative. In another series of experiments inoculations were made with placental tissue taken from the tubercular animal, with the same results. In a third series fragments of organs or sputum of the infected animal was employed, and all animals thus inoculated died of tuberculosis. These experiments led Vignal to assert that tuberculosis is very seldom congenital, and that when it attacks infants it is usually the result of post-natal infection. In the discussion that followed the reading of this paper similar views were expressed by Landouzy, Hutinel, and Bernheim. At the same time Jacobi mentioned the fact that he published the first case of congenital tuberculosis twenty years before Koch discovered the microbic origin of this disease. The experiments of Vignal do not negative the assumption that the placental circulation and the site of the placenta after its detachment are factors largely responsible for the frequent occurrence of tubercular endometritis. It has been shown that pathological conditions of the placenta increase the risks of infection of the fetus and the uterus from microbes floating in the circulation of the mother. Slight detachment of the placenta during gestation and disease of the uterus or the placental tissue itself must act as potent localizing causes; and it is in such cases that we would expect transmission of microbic diseases from mother to fetus, and disease of the puerperal uterus. Although such untoward results are most likely to happen in the case of phthisical mothers, they may occur in mothers free from pulmonary tuberculosis, provided a sufficient number of tubercle bacilli are present in the circulating blood, and locate at the placental attachment in sufficient quantity to provoke a tubercular inflammation under the conditions furnished by the puerperal uterus. The experiments of Gärtner[1] give strong proof of the correctness of the assertion made. He studied the direct transmission of the bacillus of tuberculosis through the placental circulation in the following manner: Tubercle bacilli were injected into the

[1] *Zeitschrift für Hygiene und Infectionskrankheiten*, 1893, B. xi.

abdominal cavities of female animals, in order to infect the ovum or to cause placental infection. Gärtner succeeded in showing that in peritoneal tuberculosis bacilli may find entrance into the fetus by this method of infection. Injection of bacilli into the genital circulation of pregnant animals produced tuberculosis in 10 per cent. of the embryos. It was also found that in mice with pulmonary tuberculosis, with or without secondary generalization, fetal infection often occurred. This was also the case in birds. In male animals tubercle bacilli were injected into the lungs, and bacilli were found subsequently in the semen 5 times in 32 examinations. In another series of experiments it was found that in spite of abundant bacilli in the sperm of animals with tubercular epididymitis, no fetal infection occurred. On the other hand, bacillary semen may undoubtedly cause genital tuberculosis in the female. Of 65 female guinea-pigs cohabiting with males whose testicles were rendered tubercular, 5 died of tuberculosis, the disease starting from the vagina. In mice, canary birds, and rabbits it was shown experimentally that tubercle bacilli quite often pass from mother to fetus, while the same was not found to hold true in reference to transmission from the male offspring. What experiments have indicated in regard to the transmission of tubercle bacilli from mother to child has been confirmed on a small scale by clinical observation.

Sarwey[1] describes a very interesting case of intra-uterine tuberculosis. The mother of the child

[1] *Archiv f. Gynäkologie*, B. 43, Heft. 1.

had born several healthy children. When several months pregnant she received a great mental shock. Shortly afterward fetal movements ceased. Her labor came on at about the eleventh month of pregnancy, resulting in the birth of a monstrosity deformed in many parts of the body. During the dissection of this specimen, when the organs of the throat were removed, tubercular foci were found in the bodies of the three upper cervical vertebræ, in which tubercle bacilli were demonstrated. No evidences of syphilis or tuberculosis of other organs could be found. The mother was healthy in every respect, but the father had suffered for a long time with cough accompanied with mucopurulent expectoration.

Loude[1] had an opportunity to study direct transmission of tuberculosis from mother to child in connection with 6 pregnant women suffering from tuberculosis during gestation. One of them gave birth to a living child, and inoculation of a guinea-pig with placental tissue gave a negative result; 4 of them gave birth to children who lived only a short time, and in the viscera of one of these, who died on the fourth day, tubercle bacilli were found; 2 cubic centimeters of blood taken immediately after birth from the umbilical vein and injected into the peritoneal cavity of a guinea-pig caused tuberculosis. In the other 3 children inoculation-experiments gave in 1 a negative result, whereas in the other 2 tuberculosis was produced. In the sixth case inoculation made with placental tissue and fetal

[1] *Revue de la Tuberculose,* 1893.

blood yielded positive results. Most authors agree that when extravasation or other pathological processes occur in the placental attachment, the direct entrance of microbes from the mother into the fetal circulation is not only a possible but a probable occurrence. Abnormality in the placental circulation must therefore be recognized as a condition that favors the transmission of micro-organisms from mother to fetus.

All of the foregoing observations and experiments furnish substantial proof that in some infectious diseases heredity is traceable to direct transmission to the fetus of the specific microbes floating in the circulation of the mother, through the thin wall that separates the maternal from the fetal blood. It is no more difficult to explain the migration of microbes through such a thin septum than their transportation from one tissue to another and from organ to organ in other parts of the body, more especially as the anatomical conditions for mural implantation in the placental circulation are most favorable for such an occurrence.[1] It requires no stretch of the imagination to conceive that in tubercular mothers, with or without demonstrable tubercular lesions, localization of tubercle bacilli would be more likely to take place on the maternal side of the placental circulation, and after delivery would manifest itself in the form of a puerperal metritis. The writer is firmly convinced that this explanation of the etiology of puerperal tubercular endometritis will be substantiated by future experimentation and clinical observation.

[1] Senn, *Surgical Bacteriology*, p. 24.

Tubercular endometritis is most frequently met with in child-bearing women; but, like tuberculosis of other portions of the genital tract, exceptions occur, as cases have been reported verging upon both extremes of life. The youngest patient observed by Kiwisch suffering from this disease was a girl fourteen years old, and the oldest patient was aged seventy-nine. Extension of the tubercular process from the genitals to the urinary organs in women is of rare occurrence as compared with men, as in women the genital tract has no connection whatever with the urinary system. For the same reason, infection of the genital organs from tubercular affections of the urinary tract seldom occurs except in cases in which a communication is established between the genital and urinary organs by the formation of a fistula. According to Oppenheim, the relative frequency of mutual infection between the urinary and genital organs in women as compared to men is as 1 to 3.

Symptoms and Diagnosis.—The symptoms of tubercular metritis are those of catarrhal endometritis plus the general symptoms caused by the primary affection in the secondary form of the disease. There are no pathognomonic symptoms, and in primary tubercular endometritis a positive diagnosis before operation is possible only by a microscopic examination of scrapings, which, however, can be obtained only after the diseased portion of the endometrium is accessible to the sharp curette. Menstruation, as a rule, is either suppressed or irregular; in some instances, however, it is profuse. Amenorrhea is usually caused by pulmonary phthisis, which so often precedes and attends the genital tuberculosis. Menorrhagia is sometimes a prominent feature of the disease. Leucorrhea is usually present, but may be entirely wanting. In several cases of tuberculosis of the cervix the profuse hemorrhage that attended the disease led to the suspicion of carcinoma.

In tuberculosis of the cavity of the uterus advanced to the stage of caseation and ulceration, the catarrhal secretion of the uterus is mixed at times with caseous material. If the ulcerated surface has become infected with pus microbes, the catarrhal secretion is mixed with pus and the discharge assumes a mucopurulent character. The uterus is enlarged from the beginning, is softer than normal, and in case the muscular wall is extensively infiltrated, or accumulation of the caseous material caused by inflammatory stenosis of the internal os has given rise to distention, the size of the organ is greatly increased.

The symptoms of uterine tuberculosis are often overshadowed by the complicating salpingitis or peritonitis. In the diagnosis it is very important to study the clinical history with care, and to determine, if possible, the source of infection. All the important organs, especially the lungs, should be subjected to a careful physical examination. The condition of the pelvic organs must be thoroughly investigated. If, besides the uterine affection, the Fallopian tubes are implicated in the chronic process, tuberculosis should be suspected, and the suspicion will be greatly increased by the existence of a chronic pelvic peritonitis. Examination of the vaginal secretions for tubercle bacilli is an important part of the examination. Derville, Jouin, and others were able to recognize the tubercular nature of the uterine affection by the aid of this diagnostic resource in cases in which the disease was limited to the uterus. Microscopic examination of the scrapings of the uterus has in a number of cases yielded results upon which a positive diagnosis could be based. In the absence of bacilli in the vaginal secretions, and when examination of uterine scrapings still leaves doubt as to the nature of the disease, inoculation-experiments with fragments of the affected endometrium may lead to results that will determine the nature of the infection. Suspicion should be aroused as to the tubercular nature of the endometritis if the husband of the patient is tubercular; or if the disease is hereditary in the patient's family; or, finally, if the patient is suffering from tuberculosis of any other part or organ of the body.

In rare cases the tubercular endometritis is associated with myoma or carcinoma of the uterus, still further complicating the diagnosis. Otto v. Franque[1] published a remarkable case of tuberculosis of the uterus complicated by carcinoma. The cervix uteri was the seat of a carcinoma. Total extirpation of the uterus was performed, when examination of the specimen revealed, besides the cervical carcinoma, extensive tuberculosis of the endometrium. The proliferating epithelial cells were transformed into the epithelioid cells of the tubercular product. Giant cells and tubercle bacilli were

---

[1] "Zur Histogenese der Uterustuberculose." *Virchow's Jahresbericht*, 1894, B. ii. p. 730.

demonstrated in the affected parts of the lining membrane of the uterus.

Prognosis.—The prognosis of uterine tuberculosis is always grave, as the disease is so often secondary to advanced pulmonary phthisis, and in primary genital tuberculosis it has usually extended to the tubes or is secondary to tubal or vaginal tuberculosis. In fibroid tuberculosis of the uterus, or of the uterus and tubes, rarely operative interference may eliminate all the infected tissue and be the means of preventing further local extension, and may protect the patient against tuberculosis in any other organ or diffuse miliary tuberculosis. If the uterine disease or its extensions have implicated the urinary organs, or if the disease is secondary to renal or vesical tuberculosis, death is sure to ensue in a short time. In cases of inoperable tuberculosis of the uterus death occurs from marasmus caused either by pulmonary phthisis or by the miliary tuberculosis attending or caused by the genital tuberculosis.

Treatment.—The treatment of uterine tuberculosis so far has been anything but satisfactory, owing to the difficulties encountered in making an early positive diagnosis, and the many complications that precede or accompany the disease. Conservative local measures are applicable only in cases in which the disease is limited to the vaginal portion of the uterus, when the parts affected can be removed with the hope of effecting a permanent cure by amputation of the cervix, by excising the ulcers, or by removing the infected tissues with the sharp spoon or destroying them by deep cauterization

with the Paquelin cautery. In all these procedures
the wound or eschar should be protected against
reinfection by iodoformization and packing with
iodoform gauze. All these procedures, with the
exception of the use of the sharp spoon, do not
apply to the treatment of tubercular endometritis.
In the treatment of this affection the sharp spoon
has a very limited field of usefulness, as it is impossible to remove with it all the infected tissues if
the disease affects, as it almost always does, the
fundus of the organ. Again, scraping is a very
uncertain surgical resource, from the fact that with
few exceptions the Fallopian tubes are affected primarily or secondarily. In the absence of pulmonary
phthisis or tuberculosis in any other organ inac-
cessible to direct effective treatment, vaginal hysterectomy is indicated in all cases in which the
patient's general condition warrants the operation
and the disease is limited to the uterus, tubes, and
ovaries. In Martin's clinic the treatment consists
of curettage followed by cauterization in cases in
which the disease is limited to the mucous membrane; vaginal hysterectomy is practised when the
disease has reached the muscularis. A limited
tubercular peritonitis does not contraindicate this
operation. The tubes and ovaries should always
be removed, as the disease has frequently extended
to these organs, and the organs themselves are
of no possible use after extirpation of the uterus.

Failure to remove the tubes in hysterectomy for
uterine tuberculosis is often followed by relapse, as
shown by a case reported by Zweifel.[1] The patient

---

[1] *Wiener klinische Wochenschrift*, 1892, Nos. 5, 6.

was married and twenty-eight years old. Her father died of pulmonary phthisis. Her husband contracted syphilis some years before marriage. The patient came for treatment with the diagnosis of carcinoma of the vaginal portion of the cervix. Examination of removed tissue showed it to be either syphilitic or tubercular. Specific treatment continued for some time proved of no value. The ulcerated surface was curetted and balsam of Peru and iodoform were applied. This treatment improved the local conditions, but the infection extended to the body of the uterus, causing free hemorrhage. The uterus was then curetted. Tubercles were found in the scrapings, but no giant cells or tubercle bacilli. The patient was lost sight of for a while, but returned on account of hemorrhage, and finally consented to a vaginal hysterectomy. She recovered from the immediate effects of the operation, but the affection soon reappeared in the wound. The local recurrence in this as in similar instances was undoubtedly from the tubercular tubes.

Vaginal hysterectomy is preferable to the abdominal route, as it is attended by less risk of peritoneal infection. In all these cases it should be taken for granted that the peritoneum is more or less affected in the vicinity of the organs removed, and for this reason the parts should be iodoformized, and vaginal drainage with iodoform gauze established and maintained for a sufficient length of time. The general treatment should be tonic and supporting, besides the prolonged internal administration of guaiacol, combined in anemic cases with the syrup of the ferric iodide.

PART VII.

TUBERCULOSIS OF THE FALLOPIAN TUBES.

New interest has been awakened in the pathology of diseases of the Fallopian tubes since gynecologists have resorted so frequently to removal of the tubes for inflammatory affections during the last ten years. The immense material furnished by operative work and post-mortem examinations during that time has not been used to greatest advantage in separating tubercular from other inflammatory affections. In this country tubercular affections of the tubes have been studied with special care from a scientific standpoint by Williams, Penrose, and Edebohls, whose writings have done much in calling attention to the frequency with which this affection is met with in operations for tubal affections. The monographs of Williams and Penrose show how easily it is to overlook the nature of the disease unless the specimens are carefully examined for histological or bacteriological evidences of tuberculosis. In many cases the nature of the affection is not even suspected before the operation, and the macroscopic inspection of the specimens removed furnishes us no reliable basis for a correct bacteriological or pathological diagnosis. A few years of such work as that done by the gynecologists mentioned, if made more general, will bring order out of chaos in the pathology of the different inflammatory affections of the tubes.

Winckel stated some time ago that the Fallopian tubes are affected alone in nearly 50 per cent. of all cases of tubercular disease of the genital organs. Heiberg[1] reports 13 cases of primary urogenital tuberculosis in women. In 10 cases the tubes (always bilateral), in 7 the uterus, and in 4 the ovaries were involved. The results of his observations have led him to believe that the uterus is always infected from the tubes. The same author also reported 22 cases of secondary tuberculosis of the female genital tract. Of these cases, the tubes were affected in 14, the uterus in 1c, and the ovaries in 3 cases. Schramm[2] published a very interesting anatomico-pathological study on tuberculosis of the Fallopian tubes, in which he states that this affection was met with among 3386 female cadavers 34 times, or about 1 per cent., and in tubercular women 4.2 per cent. In 27 cases both tubes were affected—five times the left tube and twice the right. The disease was found most frequently in women from twenty to forty years of age—that is, during the childbearing period.

We shall find, on further investigation, that the tubercular affection of the mucous membrane of the tubes is prone to extend over the entire surface,

[1] *British Med. Journal*, March 5, 1893.
[2] "Zur Kenntniss der Eileitertuberculose vom pathologisch-anatomischen Standpunkte." *Archiv f. Gynäkologie*, B. xix. p. 416.

frequently with extension from the tubes to the uterus on one side and to the peritoneal cavity on the other, thus constituting a pathological bridge between these two cavities. Clinical experience and pathological investigations have established the fact that the disease is generally bilateral, both of the tubes being involved simultaneously or in rapid succession. That the latter event is not uncommon is shown in many specimens by the contents of the tube and by the condition of the mucous membrane and the tubal wall, showing that this disease in one of the tubes was of longer standing than in the other. Whether the bilateral disease is the result of a common source of infection for both the tubes, or whether the opposite tube is reached by an extension of the inflammation from the tube first infected to the uterus and thence to the opposite tube, is not always easy to determine.

Werth[1] recognizes an acute and a chronic variety of tubercular salpingitis. In the former variety both the muscular and mucous coats undergo degeneration, numerous bacilli being found in the interior of the tube; while in the latter variety the tubal wall undergoes hypertrophy and cell-infiltration, and in its contents are found only a few bacilli.

Acute tuberculosis of the tubes has rarely been noted. Wernich's case,[2] in which the lungs were secondarily affected, is almost unique. Clinically the disease presents itself as a chronic bilateral salpingitis with occasional attacks of acute exacerbation,

---

[1] "Ueber Genitaltuberculose." *Centralblatt f. Gynäkologie*, July 20, 1890.
[2] "Präparat von doppelreitiger Tubentuberculose." *Berliner Gesellschaft f. Gynäkologie*, Bd. i. p. 49.

as is common to all tubercular affections, although perhaps more marked here, owing in many cases to repeated attacks of circumscribed plastic peritonitis, the most common complication of this disease.

Very little was known about the clinical aspects of tuberculosis of the tubes twenty years ago, although pathologists before that time had given excellent descriptions of the gross pathological conditions of this disease as revealed by post-mortem examinations. Beautiful colored illustrations of tubercular tubes can be found in the classical works of Cruveilhier[1] and Hope.[2] The available literature on the pathology and etiology of tubal tuberculosis is of recent date.

Pathology.—Hegar states that tuberculosis of the tubes and of the tissues in their immediate vicinity presents itself in two distinct forms: 1. The disease appears as a swelling on the side of the uterus; this swelling seldom exceeds the size of a goose's egg, can be distinctly outlined, and assumes the characteristic shape of tubal swellings; 2. The enlarged tubes are incorporated in a dense mass of plastic exudate in such a way that they cannot be outlined by palpation. This division of tubal tuberculosis might answer the requirements of the clinician, but it lacks the modern pathological foundation.

Williams[3] classifies tubercular affections of the tubes and uterus as follows: 1. Miliary tuberculosis; 2. Chronic diffuse tuberculosis; 3. Chronic

---

[1] *Anatomie pathologique*, vol. i.

[2] *Principles and Illustrations of Morbid Anatomy*, etc., London, 1834.

[3] "Tuberculosis of the Female Generative Organs." *Johns Hopkins Hospital Reports*, vol. iii. Nos. 1, 2, 3.

FIG. 12.—A Fallopian tube distended with tubercular matter: *a*, fimbriated extremity; *b*, middle; *c*, uterine extremity (Hope).

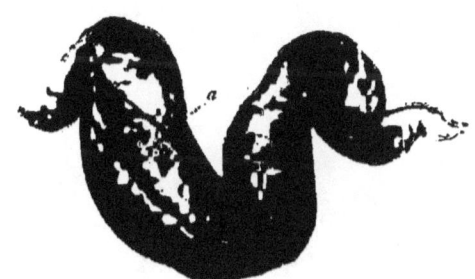

FIG. 13.—The Fallopian tube of the opposite side, cut open to show the tubercular matter (Hope).

fibroid tuberculosis. He was the first one to describe the last variety.

Miliary tuberculosis is rare as compared with the other two varieties. The process is limited exclusively to the mucous membrane, constituting simply a tubercular salpingitis. The mucous membrane is diffusely infiltrated with minute tubercles, only a few of which have advanced to the stage of caseation. The epithelial layer over the tubercles is intact, the inflammatory product being limited to the vascular portion of the mucous membrane. If the mucous membrane is the seat of active inflammatory processes, the pathological conditions and clinical evidences resemble those of catarrhal salpingitis. In the remaining forms this stage of the tubercular process is followed by successive additional pathological processes which characterize them clinically and pathologically.

In chronic diffuse tuberculosis the mucous membrane undergoes further textural changes, and the tubercular product shows more advanced degenerative alterations in the form of extensive caseation, destruction of the epithelial lining, and extension of the tubercular process to the muscular coat of the tube. This is the form of salpingitis in which the tubes are found distended to varying degrees by a cheesy material, the tube being transformed into a tubercular abscess (see Figs. 12 and 13).

Chronic fibroid tuberculosis of the tube is described by Williams as follows: "It differs from the other forms of tuberculosis in the excessive formation of fibrous tissue in and between the tubercles. Sections show the lumen greatly distended, and a few miliary tubercles scattered through the mucosa. There may or may not be accompanying inflammatory changes, the main change consisting in the excessive development of fibrous tissue both within and about the tubercle, and the relative absence of caseation. The marked feature of this form of tuberculosis appears to be its chronicity." It is apparent that Hegar's classification includes the last two forms as given by Williams.

Diffuse miliary tuberculosis is a subject of greater interest to the pathologist than the surgeon, as patients suffering from this form of tuberculosis succumb in a short time to the diffuse general miliary tuberculosis caused by tuberculosis in another organ, usually the lungs.

Tubercular pyosalpinx is a chronic affection, and may require surgical treatment in well-selected cases. Chronic fibroid tuberculosis is the most benign form of tubal tuberculosis, as the massive connective-tissue formation within the structure of the wall of the tube and around it retards or arrests the further extension of the disease.

The following is a description of the structure of the tube and the histological appearances of the tubercle tissue in the series of cases (Case VI.) reported by Williams in his monograph cited above:

"Sections through the median part of the right tube show that its lumen is much distorted and that a few miliary tubercles are scattered through the mucosa. The epithelium is preserved in most places, but the stroma is infiltrated with masses of small round and epithelioid cells, even where there are no well-marked tubercles. Many of the tubercles present a great amount of fibrous tissue both within and surrounding them, and very few centers of caseation are seen. Accordingly, we may designate the tuberculosis as of the fibroid variety and as a chronic process. The lumen is greatly constricted, apparently by the contraction of the surrounding tissues. There is no evidence of purulent or catarrhal inflammation, and no exudation within the tube. This is seen even more strikingly in the left tube. The cyst at the end of the tube is buried by flattened epithelium, and here and there in its walls are seen miliary tubercles. Sections from the superior part of the ovarian cyst show that it is buried by a single layer of flat, non-ciliated epithelium, the tissues beneath it being poor in cells, beyond which is found the typical structure of the mesosalpinx. At one point on its superior surface, just adjoining the tube and below the peritoneum, is found a small mass of tubercles in the mesosalpinx. The lumen of the left tube has lost its characteristic arborescent appearance and given place to a sieve-like structure, apparently composed of small cysts, similar to the conditions found in the so-called follicular salpingitis. None of the mucous membrane shows any sign of inflammation, but at several points typical tubercles are seen under the epithelium; none in the muscular wall of the tube. On the surface is a marked perisalpingitis, but no trace of tuberculosis is seen. The cystic condition at the end of the tube is only an exaggeration of the condition in the rest of the tube, except that no tubercles can be found in it. On the surface of the ovary are numerous adhesions, but it contains nothing else of pathological interest. No bacilli could be found in this specimen, even after hours of search. This must likewise be regarded as a primary process, and certainly a very chronic one. Its tubercular nature was not suspected until examined with the microscope."

A number of such cases have been reported by Williams and Penrose. This form of tuberculosis has its analogue in many other organs, notably the lungs, joints, and peritoneum. It is a mild infection, in which, owing either to a comparatively small number of microbes, to microbes mitigated in their virulence, or, lastly, to a greater degree of resisting power inherent in the tissues or the entire organism, the inflammatory process that is provoked assumes a plastic type. The pre-existing connective tissue and endothelial cells acted upon by the microbic cause assume active tissue-proliferation, circumscribing and to a certain degree walling off the tubercular area. The cirrhosis of the tubal wall thus produced starves out, as it were, the tubercle tissue, and, the toxins of tubercle bacilli not being present in quantity sufficient to cause coagulation-necrosis and caseation, these retrograde changes do not affect the cellular elements of the tubercles, or at any rate only to a slight extent; hence the permanency of the inflammatory product.

The different varieties of tubercular inflammation of the tubes are therefore largely determined by the intensity of the infection and the degree of susceptibility of the individual to the action of the tubercle bacilli.

The peritoneal end of the tube is almost always affected first—a fact that has been taken as proof by many that infection usually takes place from the peritoneal cavity. During the course of the disease the peritoneal ostium usually becomes obliterated, intercepting the direct communication between the peritoneal cavity and the genital tract. The disease sometimes extends uniformly over the entire surface of the mucous membrane, and, if it assumes the caseous form, results in distention and elongation of the tube, which in consequence becomes tortuous. In most instances, however, the disease remains limited to the mucous membrane, and when it reaches the uterine end results in stenosis or complete obliteration of the lumen of the tube, converting it into a closed cavity in which retention of the tubercular product necessarily takes place; this satisfactorily accounts for the accumulation of the tubercular material in the interior of the tube and for the marked hypertrophy of the muscular wall, constituting what is known as a tubercular pyosalpinx. The tubes descend behind the uterus, and, if distended throughout, form large, hard, sausage-shaped swellings. In exceptional cases the tubes are converted into large sacs. Thus, Aran describes a case that came under his observation in which the intratubal tubercular abscess ascended as far as the umbilicus; and Werth operated successfully upon a case in which the sac contained two quarts of tubercular pus. The caseous material that accumulates in the lumen of the tubes remains firm, undergoes liquefaction, or in rare cases is displaced by calcareous material.

Calcification of the tubercular product has been seen by Rokitansky, Klob, Kiwisch, Geil, and Penrose. The substitution of calcareous matter for the tubercular product is one of nature's ways in which to retard or completely arrest the disease. Such efforts are most frequently seen in the lungs. In partial calcification the disease progresses, uninfluenced by this process, in other parts of the tube, as in one of the cases reported by Penrose.[1] In one tube he found a hard mass, evidently in the tube-lumen at the junction of the middle and distal thirds. On section the lumen was found obliterated in the proximal third, and enlarged (measuring 0.9 to 1 centimeter in diameter) in the distal two-thirds. The hard body felt through the tube-wall was a calcified mass composed of five plates that measured from 0.2 to 1 centimeter in diameter. On both sides of the calcareous depot tubercular material was found. The opposite tube was the seat of similar although less advanced calcareous degeneration.

In some instances the tubercular process is limited to certain portions of the tube, the intervening part of the lumen of the tube being obliterated by cicatricial contraction. In this respect the disease bears a strong resemblance to a similar condition found in comparatively mild cases of joint-tuberculosis. Penrose describes a case in which a tubercular endometritis was complicated by such a condition of the tubes. Both tubes were the seat of two distinct tubercular nodules.

It is a well-known fact that in the vicinity of a tubercular focus with mild infective qualities there

---

[1] "Tuberculosis of the Fallopian Tubes." *Am. Journ. Med. Sciences*, March, 1896.

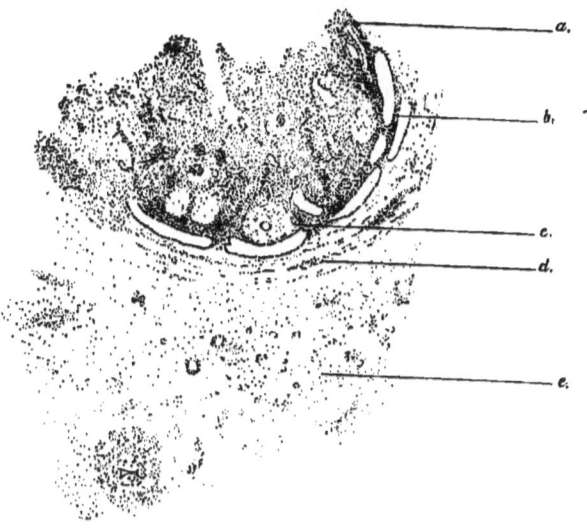

Fig. 14.—Tuberculosis of Fallopian tube, transverse section; × 50: *a*, necrotic mucous membrane; *b*, tubules; *c*, tubercle with giant cell; *d*, muscular layer, leucocytic infiltration; *e*, submucous connective tissue.

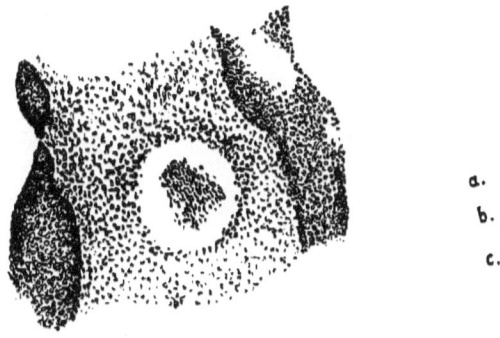

Fig. 15.—Histology of Fallopian tuberculosis; × 250: *a*, mucous membrane of lumen of tube; *b*, giant cell with numerous bacilli; *c*, zone of epithelioid and small round cells.

is established a fibrinoplastic process that results in the formation of an abundance of connective tissue, which forms a limiting wall around the infected tissues and answers an exceedingly useful purpose in retarding or arresting the further extension of the infective process. In the Fallopian tubes such a limiting process results in stenosis or obliteration of the lumen of the tube on both sides of the infected area, thickening of the muscular wall of the tube, and firm adhesions on the peritoneal side.

In cases of extensive caseation of the tubercular contents of the tubes more or less destruction of the mucous membrane is inevitable. In advanced cases the whole of the interior of the tube is denuded, and on wiping away the caseous membrane or deposit the tubercular granulations are exposed (see Figs. 14 and 15). It is in such cases that the remaining tunics of the tubes are penetrated by the tubercular infection; and perforation of a tubercular pyosalpinx has been known.

The contents of tubercular tubes are, of course, variously modified by mixed infection which not infrequently complicates the tubercular process. Mixed infection with the gonococcus, the ordinary pus microbes, and the bacillus coli commune is of most frequent occurrence, and in such cases the inflammatory product of the additional infective process forms a component part of the contents of the tubercular abscess.

Circumscribed plastic peritonitis is invariably present in all advanced cases of tuberculosis of the Fallopian tubes. The etiology of the complicating peritonitis in primary tubal tuberculosis is not easily explained. It is the result of an active reparative process in close proximity to a tubercular focus, and very much resembles in its etiology and pathology a similar form of circumscribed peritonitis so frequently found in cases of catarrhal and ulcerative appendicitis. In progressive cases of primary tubal tuberculosis the disease is prone to extend on the one side to the peritoneum and on the other to the uterus. In such cases the lymphatic glands become the seat of regional infection, while extension to the urinary organs is a very exceptional complication. Interesting is Scanzoni's case, in which the right tube was the seat of the disease, and the uterine cavity was affected on the opposite side.

Etiology.—The consensus of opinion of nearly all pathologists and gynecologists is to the effect that the Fallopian tubes are the most frequent seat of tuberculosis of the genital tract. E. G. Orthmann[1] is of the belief that tuberculosis of the female genital organs has its origin most frequently in the Fallopian tubes, basing his opinion on two cases observed in Martin's clinic. In 18 per cent. of all cases of genital tuberculosis the Fallopian tube is the primary starting-point of the disease. As the most important etiological factors he regards sexual intercourse, childbirth, and the puerperium. The disease extends from the tubes most frequently to the peritoneum, then to the uterus, and lastly to the ovaries. Ascites is present in about 15 per cent. of the cases. Orthmann favors early operative removal of the tubes.

[1] "Beitrag zur Kenntniss der primären Eileitertuberculose." *Monatsschrift für Geburtshülfe u. Gynäkologie*, 1895.

This view that tubercular salpingitis is the most frequent of all tubercular affections of the female genital organs has been greatly strengthened by the recent researches of Williams and Penrose, who have shown that it often requires a most painstaking examination of the specimens removed to demonstrate the tubercular nature of the tubal affection. These authors have shown more conclusively than was ever done before that the tubes are frequently the primary and sole seat of the tubercular process. The avenues of infection may be—1. Through the blood; 2. From the peritoneum; 3. From the uterus. Infection through the blood-supply may take place with or without an antecedent tubercular focus in another organ. Isolated tuberculosis of the tubes may occur in the same manner as joint-tuberculosis, without a demonstrable tubercular lesion in any other part of the body, by tubercle bacilli floating in the circulation becoming arrested in the capillary vessels in some part of the tube predisposed to the localization of microbes. As in the majority of cases the ampulla of the tube is first affected, it is reasonable to assume that the blood-vessels in this, the most vascular structure of the tube, present conditions predisposing that part of the tube to tubercular infection. It would certainly seem, in cases in which the peritoneum is intact and no distant organ is the seat of tuberculosis, that this explanation is the most plausible one. Another explanation, perhaps less convincing, could be offered to account for the same condition, allusion to which has already been made in the introductory remarks. Tubercle bacilli introduced from without might reach the tubes without causing any lesions of the lower part of the genital tract, and, finding here anatomical peculiarities or acquired predisposing causes, localize and produce the disease. The nearly horizontal position of the tubes and the greater liability to retention of the secretions would certainly appear to be conditions that would predispose the tubes to infective processes of a chronic character. Probably the first effect of an infection in this manner, after the bacilli have reached the fimbriæ, would be the causation of a very circumscribed plastic peritonitis sufficient in degree and extent to close the peritoneal end, after which the tubercular process would be limited to the interior of the tube, and if extension occurred, it would be in a downward direction, toward the uterus.

That tubercular peritonitis often precedes a similar affection of the tubes, and that tubercular salpingitis is frequently followed by peritoneal tuberculosis, are established and fully-recognized facts in pathology. Some of the best pathologists—such men as Virchow, Rokitansky, Klebs, and Cohnheim—believed that the genital organs were always infected by tuberculosis from the peritoneum. Schramm[1] found that in 34 cases of genital tuberculosis the peritoneum was affected in 21; Oppenheim, in 21 out of 23 cases; Osler's observations led him to say that the tubes are affected in from 30 to 40 per cent. of all cases of tubercular peritonitis; while Kaulich believes that this is the case in 50 per cent.

---

[1] "Zur Kenntniss der Eileitertuberculose vom pathologisch-anatomischen Standpunkte." *Archiv f. Gynäkologie*, B. xix. p. 417.

That tuberculosis frequently extends from peritoneum to tubes is substantiated by clinical observations as well as by experimentation. Weigert[1] has shown that in incipient tuberculosis of the peritoneum the process begins in the pelvic cavity. From the proximity of the fimbriæ to the primary tubercular foci, it would not be difficult for the bacilli to find their way into the tubes. The writer had recently an opportunity to operate on a case of tubercular peritonitis soon after the first symptoms of hydrops appeared, and found the pelvic floor studded with tubercles. One of the tubes, adherent and enlarged, was removed with the corresponding ovary. In the broad ligament were many softened tubercles. The fimbriated extremity of the tube was closed. The mucous membrane some distance from the closed end was infiltrated with minute gray tubercles. There can be absolutely no question that in this case infection of the tube was secondary to the tubercular peritonitis.

One of the most conclusive demonstrations to show the ease with which minute bodies are conveyed from the peritoneal cavity to the tubes and the uterus has been made by Pinner, who introduced powdered cinnabar into the peritoneal cavity of rabbits and dogs, and in a short time demonstrated its presence in the tubes, uterus, and vagina.

Transportation of tubercle bacilli from the peritoneal cavity into the tubes sufficient in quantity to produce the disease may occur without an ante-

---

[1] " Die Wege des Tuberkelgiftes zu den serösen Häuten." *Deutsche med. Wochenschrift*, 1883, Nos. 31 u. 32.

cedent peritonitis of the tubes to furnish the local conditions necessary for the pathogenic action of the bacilli. A case of this kind is reported from Weigert's laboratory by Curt Jani.

Ascending tuberculosis from the tubercular endometrium undoubtedly occurs in the same manner as in tuberculosis of the urinary organs—by continuity of surface—thus reaching the tubes from below. This manner of infection has been discussed more fully in connection with tuberculosis of the uterus and vagina.

In some rare cases a tube may become infected by its lumen being brought in direct communication with an adjacent tubercular lesion. In two of Mosler's cases the fimbriated end of the tube was adherent to the rectum, and by perforation of tubercular rectal ulcers direct infection followed. Such an accident might also occur in connection with tubercular ulcers of the small intestine under similar circumstances.

Among the causes predisposing to tuberculosis of the tubes must be mentioned age, the puerperium, and antecedent inflammatory affections. Although no age appears to be entirely exempt, the disease is found most frequently in women from twenty to forty years of age—that is, during the age of greatest sexual activity. Morton[1] reports a case of tuberculosis of the tubes, uterus, and vagina in a child. Doran observed that tuberculosis of these parts in children frequently appears to follow vulvovaginitis. The gonococcus destroys the epithelium of the tubes

[1] *Brit. Med. Journ.*, May 6, 1894.

and prepares the parts for tubercular infection. The puerperal uterus is more susceptible to tubercular infection than the organ in its quiescent state, and it is in such cases that the tubercular process starts at the placental site, and the tubes become infected by an ascending process.

It would be interesting to know in what percentage of cases the tubercular tubes are the seat of a mixed infection, and for this purpose the contents of the tube should be subjected to a careful bacteriological examination. There can be but little doubt that catarrhal salpingitis acts as a potent predisposing cause to tubercular infection. It is well known that in men gonorrheal affections of any part of the genito-urinary organs are a frequent and important element in the etiology of genito-urinary tuberculosis, and it is logical to assume that the same holds true in the etiology of tuberculosis of the female generative organs.

Diagnosis.—The diagnosis of tubal tuberculosis is always difficult and often impossible. The coexistence of tuberculosis in other organs should always arouse suspicion as to the tubercular nature of an existing salpingitis. If the husband of the patient is afflicted with pulmonary phthisis or with tuberculosis of the genito-urinary organs, the possibility of the tubal affection being of a tubercular character must be seriously entertained. The onset and course of the affection must also be carefully considered. The disease, if tubercular, has probably developed very insidiously. The pain and tenderness are usually less marked than in pyosalpinx. In the differential diagnosis between pyosalpinx and tubal tuberculosis Hegar places great stress on the condition of the middle third of the tube, which, he asserts, is usually free from circumscribed swellings in the former affection, whereas in the latter such swellings are found here and there in the interstitial portion of the tube. He also calls attention to the greater firmness of tubercular swellings than of ordinary pyosalpinx.

It is now well known that in the absence of tubercular peritonitis there are no characteristic conditions revealed by bimanual palpation in tubercular salpingitis that may not be simulated by ordinary catarrhal or suppurative salpingitis. Occasionally the condition of the ligaments and the lymphatics furnishes valuable diagnostic information. In some cases tubercular deposits can be felt upon the pelvic peritoneum. The existence of tubercular peritonitis, with or without ascites, is almost a sure indication of the tubercular nature of the tubal affection.

The older Chiari taught that under favorable circumstances, by vaginal or rectal palpation, the tubercular tubes could be felt as hard, tortuous, elongated swellings extending from the uterus toward the ovary, the swelling increasing in size in the direction of the ampulla. As these conditions are not always present in tubal tuberculosis, and as they may be present in hydrosalpinx and pyosalpinx, a positive diagnosis is seldom possible without additional indications of the tubercular nature of the tubal affection.

Microscopic examination of the secretions from the uterus for bacilli is an important diagnostic resource, and, although in the majority of cases of tubal tuberculosis it will prove negative, it should not be neglected in doubtful cases. In one case of tubercular pyosalpinx Edebohls made a positive diagnosis by the use of the exploring syringe. The abscess was punctured from the vagina, and in the pus removed tubercle bacilli were found.

Exploratory puncture as a diagnostic aid would, of course, be applicable only in the exceptional cases in which the tubes have become distended into large pus-sacs.

The systematic employment of the clinical thermometer the writer regards as of the greatest utility in the diagnosis of tubal tuberculosis. A slight rise in the evening temperature with a normal or subnormal morning temperature argues very strongly in favor of the tubercular nature of the affection.

Treatment.—From what has been said on the diagnosis of tubal tuberculosis it is evident that in the cases in which the disease is found affecting the tubes as a primary isolated affection, and the local and general conditions of the patient are such as to warrant surgical intervention, the diagnosis is often very doubtful, frequently remains so during the operation, and is finally made in the laboratory. In secondary tuberculosis of the tubes, except in cases in which the disease is caused by an adjacent accessible, removable, or curable focus, radical measures are, as a rule, contraindicated. The successful surgical treatment of limited primary tubal tuberculosis is an achievement of the past few years. As late as 1881, Gehle,[1] after a careful clinical and pathological study

---

[1] "Ueber die primäre Tuberculose der weiblichen Genitalien." Dissertation, Heidelberg, 1881.

of tubal tuberculosis, came to the conclusion that the removal of ovaries and tubes was absolutely contraindicated. Considering the views that were at that time entertained respecting the etiology of genital tuberculosis, and the uncertainty of diagnosis, such a verdict was sustained by most operators until quite recently. The operative treatment of genital tuberculosis in appropriate cases was strongly urged by Hegar. He emphasizes the importance of prophylactic measures, and insists that the extirpation of the ovaries and tubes, and even of the uterus, is indicated, in cases of primary isolated affections of these organs, as soon as the disease manifests progressive tendencies. Operative treatment is also indicated in secondary non-isolated affections when the changes in other organs will not produce harmful results. The contraindications are—Advanced disease of other organs; great weakness; firm and extensive adhesions of the diseased tubes to surrounding tissues and organs. In such cases opening of abscesses and cavities becomes necessary.

Werth[1] does not agree with Hegar as to the treatment of tubercular salpingitis. Hegar advises extirpation of the tubercular tubes, even when the peritoneum is affected (when the disease is limited), when the tubes are evidently the original foci, especially if they contain pus. Under these circumstances Werth simply evacuates the contents of the tube, which does not refill.

[1] *Op. cit.*

Wiedow[1] advises evacuation of the tube before excising it, in order to prevent peritoneal contamination. In cases in which the removal of firmly adherent tubes gives rise to troublesome hemorrhage he recommends incision of Douglas's pouch, packing of the small pelvis with strips of iodoform gauze, and vaginal drainage. Vaginal operations are advisable when the uterus, alone or together with the tubes, is affected.

Doran[2] says that if upon opening the abdomen it is apparent that the disease is tubercular and involves the peritoneum, investing the ovaries and tubes, the correct treatment is to close the wound at once. If cheesy fluid is present, it must be allowed to escape, and an emulsion poured in consisting of iodoform (fine crystals) 10 parts, alcohol in quantity to make it damp, distilled water 20 parts, and glycerin 70 parts. The iodoform and alcohol are rubbed together in a sterilized mortar, using the pestle well, and the glycerin is added gradually.

That operative intervention in genital tuberculosis is occasionally followed by miliary tuberculosis, in the same manner and for the same reasons as after arthrectomy or resection of tubercular joints, is shown by a case reported by Rinaud,[3] in which the removal of tubercular appendages was followed by tubercular meningitis. The operation was performed on May 5th, and on the 26th the patient was taken with vomiting, anorexia, and rise of tempera-

---

[1] "Die operative Behandlung der Genitaltuberculose." *Centralblatt f. Gynäkologie*, 1885, No 36.
[2] *Brit. Med. Journ.*, Oct. 21, 1893.
[3] *Brit. Med. Journ.*, Oct. 18, 1893.

ture. Symptoms of tubercular meningitis slowly developed, and on June 2d there was left hemiplegia and the right side of the face was paralyzed, the patient dying on the 8th. An area of brain-tissue covered with greenish exudate, and tubercular deposits along the vessels of the pia mater, were found.

In the absence of serious pulmonary complications it is the duty of the surgeon to resort to direct surgical intervention in all cases of tubal tuberculosis, with or without tubercular peritonitis. In cases of tubercular abscess of the tubes without hydrops of the peritoneal cavity, the writer would have greater confidence in tapping the abscess through the vagina (which can always be done without danger, as the tube is invariably found firmly attached to the pelvic floor) and evacuation of the contents of the abscess, followed by the injection of 10 per cent. iodoform-glycerin emulsion in doses of 2 to 4 drams. This treatment has yielded such excellent results in the management of tubercular joints that he is anxious that it should receive a thorough trial in the treatment of genital tuberculosis in well-selected cases.

If the tubal tuberculosis is attended or preceded by tubercular hydrops of the peritoneal cavity, in the absence of positive contraindications the best treatment undoubtedly is to open the abdominal cavity. The course to be pursued by the surgeon must be determined by the pathological conditions that present themselves. If the disease has originated in the tubes and has not involved the peritoneum, the tubes and ovaries should be removed whenever possible. If the primary tubal disease has given rise to limited tubercular peritonitis, the same course should be pursued. In such cases it is the universal practice of the writer to iodoformize the peritoneal cavity by pouring into it from 2 to 4 drams of a 10 per cent. iodoform-glycerin emulsion, and to establish drainage by inserting into the lower angle of the incision a Keith's glass drain loosely packed with iodoform gauze. The drainage is continued until all secretions have ceased, when the drain is removed and the wound is closed by secondary sutures inserted at the time of operation. If the tubes are not much distended, but are so firmly imbedded in strong and extensive adhesions as to render their removal a dangerous procedure, they should be left, and the peritoneal cavity treated as advised above. The author is fully satisfied that in a number of his cases thus treated a permanent cure was effected. He has also, in a number of cases of tubercular hydrops of the peritoneal cavity complicated by tubal disease, resorted to tapping and iodoform injections, with the most satisfactory results. In two cases treated by laparotomy and drainage the ascites returned after the wound healed, and yielded later to this method of treatment. If the tubes are found much distended and so firmly attached as to render their extirpation impracticable, they should be incised, evacuated, scraped, and packed with a strip of iodoform gauze which is brought out at the lower angle of the wound on the side of the glass drain. If the uterus is involved primarily or secondarily, and the patient's general condition warrant operative procedure, vaginal hysterectomy with removal of the adnexa is indicated. If the peritoneum is found affected, the parts within reach should be iodoformized and vaginal drainage established with iodoform gauze.

In conclusion, it may be stated that tubal tuberculosis should be treated upon the same general surgical principles as similar affections in any other part of the body.

PART VIII.

TUBERCULOSIS OF THE OVARY.

THE ovary, the essential organ of generation in the female, is seldom the seat of isolated primary tuberculosis, but is quite frequently implicated secondarily in cases of peritoneal and tubal tuberculosis. Klob, Rokitansky, and Virchow believed that the ovary was seldom, if ever, the primary seat of infection. B. Wolff[1] mentions the case observed by Scanzoni in 1844 as the first instance of tuberculosis of the ovary, and, as the second, the one reported by Négrié in 1863; in the latter case the diagnosis was confirmed by microscopic examination. Since that time isolated cases of caseous ovaritis have been described by different authors, but the subject did not receive much attention until Terillon and Klebs emphasized its importance. The former author in 1887 stated that in 6 cases of genital tuberculosis the ovaries were affected in 3; and Klebs[2] as early as 1876 speaks of 3 or 4 cases which he had an opportunity to examine and which were usually complicated by peri-oöphoritis. As compared

---

[1] "Aus dem Tenkenberg'schen pathologisch-anatomischen Institut in Frankfurt a. M. 1896." *Ueber Tuberculose des Eierstockes.*

[2] *Handbuch der pathologischen Anatomie*, B. ii. p. 825.

with tuberculosis of the tubes and uterus, Klebs regarded tubercular ovaritis as of very rare occurrence. What attracted his attention at the time was that in his cases the tubes were not affected, and he consequently attributed the disease to blood-infection. Similar cases were reported by Pillaud and Späth. In one of Gusserow's[1] cases both ovaries were enlarged to the size of a hen's egg, and in their interior were found numerous tubercles and cheesy masses. In the second case the ovaries presented a similar appearance, but had attained only the size of a pigeon's egg. In both cases the ovarian disease was complicated by tubercular endometritis. Heiberg has seen secondary tuberculosis of the ovary in children under eleven years of age. Wagner[2] found tubercles in the ovaries of 5 per cent. of tubercular women. Scanzoni in the case referred to found in the ovary numerous tubercles the size of a millet-seed, partly gray and partly yellow, and in some places beginning softening. Rokitansky saw in one case, besides tuberculosis of the lungs, intestines, peritoneum, and tubes, the left ovary infiltrated by numerous cheesy tubercles, those in the periphery involving its fibrous investment. In the right ovary the disease was much less advanced. The collection of 10 cases by Talamon[3] of tuberculosis of the female genital organs in girls less than fifteen years of age shows that in 9 cases the uterus was affected and in 5 the ovaries.

[1] " De muliebrium genitalium tuberculosi." Dissertation, Berlin, 1859.
[2] *London Lancet*, Dec. 27, 1889.
[3] "Pelviperitonité chez une petite fille de 6 ans. Tubercules des ovaires. Metrite suppurée enkystée. Méningite tuberculeuse." *Le Progrès méd.*, 1878, p. 983.

From the cases that have been reported and from the results of post-mortem examination it appears that tuberculosis of the ovary as a primary lesion is quite rare; on the other hand, the ovary is frequently implicated in tuberculosis of the peritoneum and of the genital tract. B. Wolff has shown that tuberculosis of the ovary very often presents itself in the post-mortem room as a bilateral affection.

Pathology.—Tuberculosis of the ovary in its pathology has much in common with the same disease in the testicle. It occurs either in the form of miliary tubercles or as a caseous inflammation, the cheesy form always being preceded by miliary infiltrations. Acconis and B. Wolff have very recently found in tubercular ovaries small cysts surrounded by a firm wall of connective tissue, which in favorable cases results in obliteration of the cavity by cicatricial contraction. Verneuil,[1] from his pathological studies of this subject, classifies tuberculosis into two varieties: 1. Diffuse tubercular infiltration of the entire organ, which he regarded as a very rare disease; 2. Encysted form, in which the foci are larger and undergo caseation, eventually terminating in the formation of tubercular abscesses, the walls of which are studded with tubercles. This classification cannot be improved upon to-day.

In cheesy ovaritis the caseous material either remains firm or liquefaction occurs, transforming the tubercular focus into a tubercular abscess. In primary tuberculosis of the ovary the miliary tubercles are subject to the same degenerative changes as

[1] *Thèse de Paris*, 1880.

in any of the other organs, as the tubercle-tissue is exposed to the same influences. Coagulation-necrosis, caseation, and liquefaction of the caseous material occur in their regular order. In secondary ovarian tuberculosis the miliary nodules are found at the post-mortem with little or no evidences of caseation, as the patients succumb to the primary disease before a sufficient time has elapsed for the tubercle-tissue to undergo these different retrograde degenerative processes.

Tubercular peri-oöphoritis resulting from peritoneal tuberculosis or from extension of the tubercular process from the tubes is of much more frequent occurrence than tuberculosis of the parenchyma of the organ. It is not always easy to determine, either at the post-mortem examination or from examination of specimens removed by operative procedure, whether the disease had its origin beneath or upon the surface of the albuginea, as in cases in which this disease had a peripheral origin the whole ovary eventually may be destroyed by the invading tubercular process, whereas, if the disease originated within the substance of the organ, perforation of the capsule results in the formation of a parovarian tubercular abscess. Except in the rare cases of blood-infection and local infection of a ruptured Graafian follicle from the peritoneum or tube, infection from the same sources would naturally at first appear in the form of a tubercular peri-oöphoritis.

The histological structure of ovarian tubercles does not differ essentially from the structure of tubercles as found in any other part of the genital organs, except that the epithelioid cells, owing to their histological origin, bear a strong resemblance to the large cells of a corpus luteum, part of the epithelioid cells being supplied by the epithelial cells that line the Graafian follicles, the remainder being furnished by the stroma of the organ. In ovarian and parovarian tubercular abscess the inner surface of the wall is studded with miliary tubercles which also penetrate the surrounding tissues; these tissues in the course of time break down, enlarging the cavity and imparting to the disease its progressive character. In extension of the disease from tube to ovary the tubercular abscess is usually located between the ovary and the fimbriated extremity of the tube, surrounded and walled off by the products of a circumscribed plastic peritonitis. Such tubo-ovarian abscesses are not infrequently found in operations for tubal tuberculosis.

Ovarian cysts occasionally furnish the necessary conditions for the localization of tubercle bacilli. The tubercular complication was never suspected before the operation, and in each instance was detected by examination of the specimen. It is seldom that ovarian cysts are examined with sufficient care to determine the existence of complications; hence tuberculosis in ovarian cysts has occurred much more frequently than the few cases on record would indicate.

Spencer Wells[1] describes a cystic tumor of the ovary in which Fox found tuberculosis. Upon the

[1] "Tubercle of the Ovary." *Transactions of the London Pathological Society*, 1864, vol. xv. p. 175.

external surface of the unilocular cyst were found, underneath the peritoneal investment and firmly attached to the surrounding tissues, numerous nodules of the size of a grain of pepper and very dense. In the center these nodules were opaque and cheesy. The nodules contained no blood-vessels. In their vicinity were delicate pseudo-membranes which were covered with miliary granulations. Similar granulations were found in the substance of the cyst-wall.

In the Pathological Institute in Prague is a specimen of an ovarian cyst, examined and prepared by Klebs, the seat of a similar tubercular complication. Infection in this case evidently took place from the tubercular uterus.

Sänger[1] describes an ovarian cyst in a woman sixty-seven years of age. The cyst had become infected with tubercular material by repeated tapping through an abdomen which was the seat of a tubercular peritonitis.

Madlener[2] mentions a case of tubercular ovarian cyst and tubercular uterine polypus. A smooth, movable cyst the size of a man's head contained tubercular material; and a not less interesting rarity was found in the fundus of the uterus in the form of an edematous polypus 4 centimeters long, tubercular, containing giant cells and tubercle bacilli.

Baumgarten[3] examined an ovarian cyst, removed from a girl fourteen years old, which showed marked

---

[1] *Brit. Med. Journ.*, Sept. 20, 1889.
[2] *Centralblatt f. Gynäkologie*, June 2, 1894.
[3] "Fall von Kystoma ovarii mit tuberculöser Entartung der Cystenwand; Extirpation; Genesung." *Virchow's Archiv*, B. xcvii. Heft 1.

evidences of tubercular infection. During the operation it was noticed that the peritoneum was covered in many places by yellowish-white nodules and was somewhat thickened and abnormally vascular. The surface of the tumor was studded with nodules varying in size from a millet-seed to that of a pea. Although no bacilli were found, the structure of the tubercles showed their tubercular character. The patient was reported to be in good health six months after the operation. It is more than probable that in some of these cases the cyst-wall was infected from the peritoneum, while in others a hematogenetic origin must be taken for granted.

Ehrendorfer[1] reports a case of tubercular ovarian cyst removed in the Innsbruck clinic. The patient was twenty-three years old. The tumor, the size of a child's head, was partly intraligamentous and partly retroligamentous. The cyst ruptured during the operation, and a yellowish-green pus escaped. The intestine was also injured during the enucleation of the cyst. The rent was at once sutured and the cavity packed with iodoform gauze (after Mikulicz). The patient died the next day. At the postmortem examination numerous tubercular nodules were found in the vicinity of the cavity. The left adnexa contained tubercles and caseous foci. The lungs and other organs were free from tuberculosis. The cyst-wall showed the usual structure of ovarian cystoma, but its inner surface was covered with a cheesy material. It is uncertain whether the tubercular process commenced in the

---

[1] "Tuberculöse Ovarial-cyste." *Wiener klin. Wochenschrift*, 1896, No. 15.

ovary or whether the ovary became infected from the tube.

It is not difficult to understand that the increased vascularity of the peritoneum and the subperitoneal tissues of the cyst-wall would determine localization of tubercle bacilli present in the pelvis, but not in sufficient quantity to cause tuberculosis of the normal peritoneum. The atypical vascularity of the tumor-tissue, on the other hand, cannot fail in increasing the liability to infection from the blood, so that cystic and other tumors of the ovary may under certain circumstances become the seat of parenchymatous or peripheral tubercular infection.

B. Wolff[1] collected 72 cases of ovarian tuberculosis, and found that the age of greatest sexual activity—that is, the time between fifteen and thirty-five years—acted as a potent predisposing cause.

Infection through the blood as the immediate cause of tubercular ovaritis is strongly emphasized by Mosler, Klebs, and Guilleman.

Etiology.—Primary tuberculosis of the ovary is caused by hematogenetic infection in the same manner as tuberculosis of the testicle. Such a mode of infection is possible without antecedent disease in any other part of the body. Undoubtedly the infection in such cases is often preceded by textural changes that determine the localization, such as small cysts and chronic inflammatory processes. Klebs believes that infection is often determined by trauma, such as rupture of a Graafian follicle,

---

[1] *Op. cit.*

the tubercle bacilli locating in the follicle that has undergone such a change. Tubercular ovaritis caused by blood-infection is often bilateral, as in the two cases reported by Gusserow. Primary tuberculosis of the ovary may also take place in the peritoneum, through a ruptured Graafian follicle, by the entrance of bacilli that have found their way into the peritoneal cavity through the rent into the follicle, resulting first in a circumscribed follicular tuberculosis which later may involve the entire organ. Primary peri-oöphoritis is probably not infrequently produced in this way, the peritoneal investment of the ovary presenting the necessary localizing conditions, which may be absent in the adjacent normal peritoneum.

Secondary tuberculosis is caused either by the transportation of tubercle bacilli from another tubercular organ through the blood-vessels or lymphatic channels, or by the direct extension of a tubercular process from an adjacent organ. In the first instance the disease of the ovary constitutes only a part of a diffuse tubercular infection, while in the latter case it usually occurs in connection with the tubal or peritoneal tuberculosis. In both cases the disease begins most frequently as a peri-oöphoritis, which eventually may implicate the entire organ. In one of the writer's cases of primary peritoneal tuberculosis one of the ovaries was the seat of well-marked peri-oöphoritis with miliary infiltration of the parenchyma of the organ. The peritoneal investment of the ovary was studded with tubercles the size of a millet-seed, which on section showed a minute central focus of caseation. The ovary was considerably enlarged; its parenchyma was softened and was diffusely infiltrated with gray tubercles of more recent date. The ovary was adherent to the parietal peritoneum and the fimbriated extremity of the tube by the recent but quite firm adhesions.

In chronic tubercular disease of the tubes a tubo-ovarian tubercular abscess is frequently developed, in which the remains of the ovary and the fimbriæ of the tube often form a part of the abscess-wall. Localized or diffuse tubercular peritonitis is always present in such cases.

Secondary tuberculosis of the ovary has been met with in children one year old, and quite frequently before the age of puberty. Hennig[1] cites a case of a child from Amboyna, twelve years old, who had tubercular disease of both ovaries, one tube, and the uterus. The greatest number of cases of ovarian tuberculosis, however, occur in women twenty-five to thirty-five years of age.

Symptoms and Diagnosis.—Isolated primary tuberculosis of the ovaries is beyond the reach of our present means of diagnosis, as it is not characterized by any symptoms that would enable us to differentiate it from other inflammatory affections. In the diagnosis of tubal and ovarian tuberculosis (the tubes and the ovaries being most frequently simultaneously affected or becoming involved successively), the clinical symptoms, and the results of examination enable us to recognize tubo-ovarian dis-

---

[1] *Die Krankheiten der Eileiter und die Tubenschwangerschaft*, Stuttgart, 1876.

ease, but in the absence of microscopic proof do not yield positive diagnostic results. The existence of tubal or ovarian disease in connection with peritoneal tuberculosis would leave a strong suspicion of the tubercular nature of the former affections. The differentiation between a tubo-ovarian tubercular abscess and a pyosalpinx is impossible, short of tapping the abscess and bacteriological examination of its contents.

Some of the important features of genital tuberculosis, wherever it may be located, are the chronicity of the disease, its painlessness as compared with other inflammatory affections, and a slight nocturnal rise of the temperature—clinical witnesses which should be accurately studied and their importance carefully weighed in the differential diagnosis between tuberculosis and other inflammatory conditions. Amenorrhea is the rule, caused by the organic affection of the ovaries and the general ill-health of the patient. Bourceret asserts that patients suffering from tubercular ovaritis are more frequently hysterical than those suffering from other forms of pelvic inflammation, but this part of the clinical picture would be of little or no importance in the differential diagnosis.

Treatment.—The intentional removal of tubercular ovaries has never been done, as a positive diagnosis is possible only after opening the peritoneal cavity. Operative treatment so far has been limited to the removal of tubercular ovaries in cases in which the ovarian disease was complicated by peritoneal or tubal tuberculosis. The removal of the diseased ovaries is indicated in all cases of uterine, tubal, or peritoneal tuberculosis that warrant operative interference, and when this can be done without adding much to the immediate risks of the operation. In cases of tubo-ovarian tubercular abscess, tapping of the abscess through the vagina and evacuation of its contents, followed by iodoformization of the cavity, may be tried in appropriate cases. In the vaginal removal of a tubercular uterus the tubes and ovaries should invariably be removed, as we have no assurance that these organs have not been reached by infection. In operations for tubercular peritonitis by laparotomy, if an ovarian or tubo-ovarian abscess is found, and examination reveals conditions that contraindicate a radical attempt, the abscess-cavity should be aspirated, incised, scraped, iodoformized, and packed with iodoform gauze which is brought out at the lower angle of the wound on the side of the tubular drain. Preliminary aspiration of the abscess is calculated to diminish the risks of a new infection of the peritoneal cavity. During the treatment of the abscess the remaining abdominal organs should be carefully protected by a gauze compress. The general treatment should be conducted on the same principles as for tuberculosis of any other organ.

PART IX.

TUBERCULOSIS OF THE BLADDER.

The diagnosis of the different inflammatory diseases of the bladder is still in its infancy. Until quite recently surgeons seldom made an attempt to differentiate between the different inflammatory affections, being satisfied to make a practical distinction only between acute and chronic inflammation, the so-called vesical catarrh. The modern scientific surgeon is no longer satisfied with such an unscientific and practically useless classification, but is desirous, by improved methods of investigation, of basing the inflammatory affections, so far as their diagnosis and treatment are concerned, upon a bacteriological groundwork.

From bacteriological and practical aspects it is very important to make a wide distinction between chronic cystitis caused by pus microbes and that caused by the bacillus of tuberculosis. The most serious mistakes in treatment have been made in cases of chronic cystitis not properly diagnosticated. The existence of a chronic inflammation of the bladder, in the absence of tangible evidences of infection from gonorrhea, chronic obstruction, or by instrumentation, should always leave a suspicion of the tubercular nature of the affection.

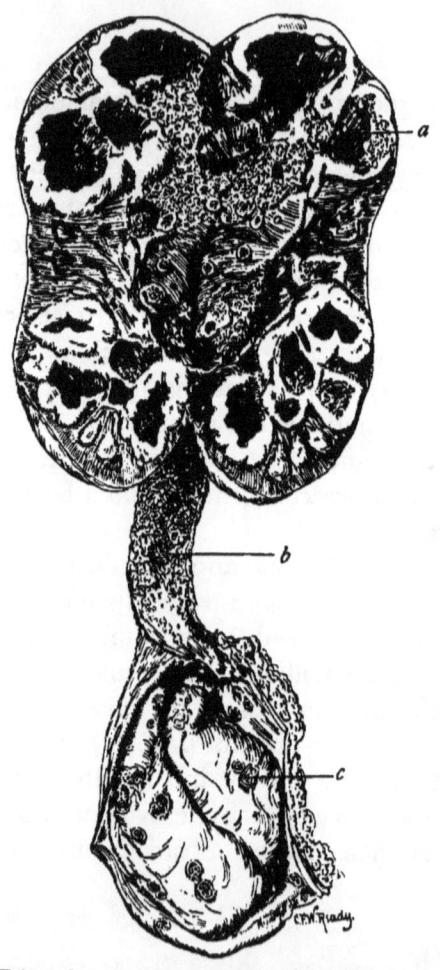

FIG. 16.—Tubercular cysto-uretero-pyelonephritis, one-third natural size: *a*, large tubercular abscesses of kidney; *b*, ureter (shortened and dilated), the mucosa nodular and thickened; *c*, tubercular ulcer of the bladder.

Pathology.—The two places in which tuberculosis of the bladder is most likely to commence are the ureteral orifices and the trigone of the bladder. The former starting-point of the disease is the rule when the bladder becomes involved by a descending tubercular ureteritis—that is, when the disease is secondary to renal tuberculosis; the trigone is usually the starting-point in primary tuberculosis of the bladder, and in men also by the extension of the disease from the genital organs. In the urinary bladder the tubercular process begins with the formation of typical gray nodules in the mucous membrane; these nodules, as in other organs, become confluent, caseate, soften, and produce ulceration. They should not be mistaken for the normal lymph-follicles that exist in the healthy mucous membrane. In more acute cases the disease leads to more diffuse cheesy infiltration and extensive ulceration (see Fig. 16).

In the chronic form of bladder-tuberculosis small groups of tubercles coalesce and form small cheesy masses, which, by caseation and softening of the cheesy material, form in a typical manner the so-called lenticular ulcers, characterized by a flat base covered with caseous material, and sharp, ragged borders, within which little tubercles can be found, some distance from the free surface, infiltrating the adjacent tissue. "The ulcers are at first circular, usually superficial; their edges are slightly raised, and surrounded by a pale anemic zone set in the deeply hyperemic mucous membrane. In the bladder these ulcers frequently coalesce, forming larger ones of variously irregular forms" (Newman). Penetration of the wall of the bladder frequently leads to the formation of perivesical abscess and fistula-formation, a part or all of the urine escaping through the fistulous opening. The violent contractions to which the walls of the bladder are subjected during the course of the disease, and the products of the chronic inflammatory process, lead, as a rule, to great thickening and progressive contraction of the lumen of the organ. The irregular contraction and dissemination of the tubercular process often result in sacculation of the organ. In one of the writer's cases the capacity of the bladder hardly exceeded a tablespoonful. In one of S. D. Gross's cases[1] the bladder was found contracted to the volume of a hen's egg and contained a few drops of a reddish urine along with small flakes of mucus. Its walls were somewhat thickened, especially in front and below. In this case the prostate, the left kidney, and the lungs were the seat of tuberculosis. In old cases the surface lesions are usually most marked in the vesical trigone. Granulations can rarely be seen, but when visible they appear like fine gray dots, sometimes confluent, but never in such masses as are seen in the kidneys. Tapret describes the ulcers as strongly resembling herpetic ulceration. In depth they vary from a very superficial surface defect to penetration of the entire thickness of the bladder-wall, which occasionally has resulted in perforation. Perforation and perivesical abscess-formation are more likely to take place if the wall of the bladder becomes deeply infiltrated and the tubercular process is rather acute and reaches the perforated surface of the bladder

[1] *A Practical Treatise on the Urinary Organs*, p. 316.

before any great thickening of its walls has occurred. Perforation into the perivesical loose connective tissue at the base of the bladder not covered by peritoneum may lead to rapidly spreading and diffuse tubercular infection, which may terminate by perforation of a tubercular abscess into the rectum or the urethra, and in women may cause a vesicovaginal fistula; or such abscess may open upon a surface of the body distant from the bladder, as the peritoneum or the groin.

The extension of the tubercular inflammation over the surface and in the direction of the different tunics of the bladder-wall is hastened in case the bladder becomes infected with pus microbes, which is so often the case, and which is so frequently caused by the needless use of instruments in the useless search for stone in the bladder, which a beginning vesical tuberculosis often mimics closely.

The complications most frequently encountered in post-mortem examinations of patients who have died of the direct or indirect effects of tuberculosis of the bladder are tuberculosis of the lungs, kidneys, genital organs, and peritoneum, and perivesical tubercular abscesses with or without fistula-formation.

Etiology.—Tuberculosis of the bladder is caused either by infection through the blood, by extension of a tubercular process by continuity of surface from the kidney or the genital organs, or by the rupture of a tubercular abscess into the bladder. Vesical tuberculosis is found more frequently in males than in females, and is a disease of early and middle life. Tuberculosis in the male is usually associated with tuberculosis of the seminal vesicles and of the prostate. Localization of the tubercle bacilli in the mucous membrane of the bladder is favored by antecedent affections of the urinary tract. Primary tuberculosis from infection through the blood is so rare that König doubts its existence. Infection takes place most frequently from the kidneys, less frequently from the prostate, seminal vesicles, and epididymis. The resistance of the mucous membrane of the bladder to tubercle bacilli is great. In many cases tuberculosis of the kidneys may exist for several years without affecting the bladder. The mucous membrane of the bladder can be irrigated with urine containing tubercle bacilli for years without becoming tubercular.

Clado[1] pointed out that tubercular granulations in the bladder do not, as is claimed by some authors, occupy the mucous tissue, but the mucous membrane itself—that is, the subepithelial layer. He believes that this is due to the presence of a well-developed capillary network in the mucous membrane, which determines localization of the bacilli floating in the general circulation. He is of the opinion that tuberculosis of the bladder is more frequently caused in this manner than by extension by contiguity from the pelvis of the kidney. Nearly one-half of all the cases of chronic cystitis in which the urine contains pus and mucus are of a tubercular nature, if these urinary conditions constitute the chief symptoms. Primary tuberculosis of the blad-

---

[1] "Note pour servir à l'étude des lésions anatomico-pathologiques de la tuberculeuse vé-icale." *Ann. des Maladies des Org. Gén.-urin.*, 1887, p. 46.

der, of hematogenic origin, is exceedingly rare. Secondary infection occurs most frequently from the prostate or the kidney, and least frequently as a result of an ascending affection of the testicle. Most frequently the infective process is continuous, extending from the pelvis of the kidney to the ureter and from the ureter to the bladder. In other cases the bladder is involved by the rupture into it of a tubercular prostatic abscess, or by extension along the ducts to the urethra and from there to the bladder. An ascending tuberculosis of the ejaculatory ducts in other cases precedes the bladder-affection. A previous gonorrheal cystitis not infrequently prepares the soil for tubercular infection. König observed a case in which a turpentine-intoxication first produced active symptoms in a case of latent catarrhal tubercular cystitis caused by a tubercular kidney.

Although no age is exempt, tuberculosis of the bladder occurs most frequently in men between seventeen and forty years of age. Baudet[1] records a case in a boy fifteen years old; in this case the earliest point of invasion, so far as could be determined, was the testicle, then the prostate and bladder, thence along the ureter to the kidney. The writer has seen, in a girl nine years of age, a case of primary vesical tuberculosis that extended to both kidneys and proved fatal in less than a year.

Symptoms and Diagnosis.—The symptoms of vesical tuberculosis often mimic those of stone in the bladder so closely that no less eminent a surgeon

[1] *Journal de Médecine de Bordeaux*, Aug., 1890.

than Dupuytren cut a boy two and a half years of age for stone in the bladder, only to find that he had made an erroneous diagnosis. The post-mortem revealed tuberculosis of the bladder. As in tuberculosis of other parts of the urinary system, the symptoms are variable and often misleading. Even the presence of tubercle bacilli in the urine, indicating as it usually does tuberculosis of the urinary system, does not locate the anatomical seat of the lesion. If the urine contains an abundance of bladder-epithelium and tubercle bacilli, the diagnosis of vesical tuberculosis is probable, but not positive. Besides a systematic study of the clinical history of the case and a careful search for a source of infection, the symptoms should be considered separately and collectively in each case before coming to any definite conclusions. The urine must be examined in every case for bacilli and histological elements, either to negative or to support a probable diagnosis. A patient suffering from tuberculosis of any other organ who develops vesical symptoms is in the majority of cases the subject of secondary tuberculosis of the bladder.

As has been stated before, tuberculosis of the bladder is extremely rare as a primary isolated affection. Bryson[1] says that of 174 cases observed by him which were positive examples of tubercular disease of the urinary organs, only 18 gave unmistakable evidences of renal involvement. He adds: "Neither in my notes nor within my recollection is there a single case where the disease was primary for

---

[1] *Journal of Cutaneous and Genito-urinary Diseases*, Nov., 1894.

the body in the kidneys, and I cannot elicit from these any figures as to the relative frequency of the primary infection of these glands in the urogenital cycle. In a majority of cases of renal tuberculosis the bladder gave the first clinical signs of mischief, and without any exception the lower urinary, and in males frequently the genital, organs were distinctly involved when first seen by me; but in a certain proportion of cases the middle lower urinary passages were infected in a manner to fully justify the belief that the disease came to them from above—*e. g.*, in those striking cases of surface infection of the vesical outlet, the trigone, and ureteral orifices, which are seen conjoined with renal tuberculosis, and which, under observation, glide steadily into unmistakable vesical tuberculosis, often spurred into activity by injudicious instrumentation." In the experience of other surgeons primary tuberculosis of the upper portion of the urinary tract is certainly much more frequent than stated by Bryson, but primary tuberculosis of the bladder is quite rare as compared with secondary tuberculosis by extension from the kidney, the genital organs, or the lower portion of the urinary passage. The disease is initiated by a frequent desire to urinate and by pain after emptying the bladder, with slight hematuria at longer or shorter intervals. Urination becomes more frequent as the disease advances, and after the neck of the bladder has been reached incontinence of urine becomes a conspicuous clinical symptom. The urine exhibits the same appearance and contains the same morphological constituents as in cases of chronic catarrh of the bladder (see Fig. 17). In the beginning of the disease it is acid and contains pus, bladder-epithelia, and a small quantity of albumin. If the kidneys are affected at the same time, the albumin is more abundant. If secondary infection with pus microbes has occurred, as is often the case, during the early stage of the disease, the urine is alkaline in reaction and often ammoniacal, and

FIG. 17.—Histological elements found in the urine in chronic inflammation of the bladder (Peyer).

then contains also a larger amount of mucus and pus-corpuscles and disintegrated red blood-corpuscles, besides the large flat epithelial cells from the bladder. As soon as the cheesy material on the surface of the bladder softens and disintegrates, fragments of detritus are found in the urine. Tubercle bacilli are not always present, and their detection is often very difficult. They are found most frequently in cases in which the process has given rise to ulceration. The disease may remain stationary for years; in other cases the progress of the disease is very rapid, and the vesical distress very great and almost continuous. Strangury is often very severe and prolonged. Perforation occurs most frequently into the rectum, but may also take place into the colon or the small intestine, and in women in the direction of the vagina.

Pain appears as one of the earliest symptoms, and increases with the progress of the disease. As compared with pain in cases of stone in the bladder, it may be stated that in tubercular disease of the bladder pain is present before, during, and after urination, but is not so much aggravated after emptying the bladder as is the case in stone in the bladder, and there is seldom more than a trace of blood mixed with the last drops of urine. In tubercular disease the pain is most acute about the middle of the act of urination, increasing toward the termination and subsiding slowly soon after the bladder is empty. The tenesmus is also less severe than in vesical calculus. Pain at the end of the penis is also less constant and severe. Sudden stoppage of the flow of the urine does not occur during urination. In renal calculus the appearance of pus is preceded by hemorrhage, which is not the case in tubercular disease. In suppurative pyelitis the purulent stage develops early, the pus being constantly present, although variable in quantity, and the general symptoms are more severe during the early stages of the disease. In cases of malignant tumors of the kidney the urine seldom contains the products of inflammation and the hematuria is more pronounced.

Tenderness.—Increase of pain on pressure over the pubes or through the rectum is present in all forms of inflammation of the bladder, but is less marked in tuberculosis than in suppurative cystitis. A sudden increase of tenderness would indicate the occurrence of a mixed infection.

Hematuria.—Blood in the urine is found in many cases of vesical tuberculosis as a prodromal symptom, but usually only at intervals and in sufficient quantity to impart a reddish tint to the urine. If the urine is bloody, evacuation of the bladder is sometimes followed by the escape of a few drops of pure blood. The hematuria is never as profuse as in tumor of the bladder, and usually is not constant. Stapfer,[1] on clinical material obtained mostly from Parisian hospitals, has made a careful study of hematuria as a diagnostic sign of vesical tuberculosis. He maintains that vesical hemorrhage is often the first and only symptom of tuberculosis of the bladder, and that it is not always attended by the remaining symptoms of a subacute cystitis. In other cases the hematuria is preceded by chronic inflammatory affections of the testicle and prostate, especially induration of the epididymis of a tubercular character. Hemorrhage after the prodromal stage indicates that the tubercles have advanced to the stage of ulceration.

Retention.—Retention of urine in the course of vesical tuberculosis means the existence of an in-

[1] " Essai de diagnostic de l' hématurie vésicale causeé par la tuberculisation." *Thèse de Paris*, 1874.

flammatory swelling of the internal urethral orifice, enlargement of a tubercular prostate, or destruction of the neck of the bladder by the tubercular process.

Incontinence.—Incontinence of urine invariably shows that the neck of the bladder has been reached by the tubercular process. It may appear during the early part of the disease if the infection starts from the trigone, or late when the disease originated in the vicinity of the orifice of the ureter.

Pus.—Pus in the urine, with or without fragments of tubercular matter, appears after the inflammatory process has reached the surface of the mucous membrane, and becomes more abundant when caseation and ulceration set in or if the bladder becomes the seat of a secondary pyogenic infection. As soon as the urine becomes alkaline it contains mucus, which is often so abundant that a copious viscid sediment adheres to the bottom of the vessel. With the appearance of pus, bladder-epithelia can be found in the urine. In mixed vesical infection pus becomes very abundant—so much so that on standing the sediment may form half the volume of urine.

Bacilli.—The tubercle bacillus may be detected in the sediment by the ordinary methods or by centrifugation, but its staining and detection are somewhat difficult. Its presence can also be determined by cultivation on artificial nutrient media and by inoculation-experiments. Strümpell[1] relies on the presence of tubercle bacilli in the urine in making a positive diagnosis of vesical tuberculosis. Not every patient whose urine contains mucus, pus, and

[1] *Münch. med. Woch.*, Nos. 30, 31, 1887.

bacilli suffers from tuberculosis of the bladder, as the bacilli may come from the kidneys or from some portion of the genital tract the seat of the disease. König is of the opinion that many cases of chronic cystitis are of a tubercular nature, although repeated examinations for bacilli have proved negative. It is probable that the staining qualities of the tubercle bacillus are diminished or entirely destroyed in alkaline and ammoniacal urine.

In conclusion, it may be stated that when the existence of a catarrhal affection of the bladder has been determined, it becomes necessary to ascertain its nature. If tuberculosis is suspected from a clinical study of the case, a careful search for additional tubercular affections is of paramount importance. Unimpaired general health does not necessarily exclude or militate against the tubercular nature of the affection. Exclusion of other causes of catarrhal cystitis, such as stricture, gleet, etc., is very important in the differential diagnosis. The two most reliable diagnostic resources are the demonstration of the presence of tubercle bacilli in the urine and the results of inoculation-experiments, but even these evidences cannot be relied upon to the exclusion of a careful study of the clinical history and symptoms. If the bacilli cannot be found, the injection of a few drops of the urine-sediment into the eye, a joint, the pleura, or the peritoneal cavity of a rabbit will often succeed in reproducing the disease, and upon the results of such experiments we must then base our diagnosis. Such experiments were made for diagnostic purposes long before the bacillus of tuberculosis was discovered. The positive results of such experiments and the detection of bacilli in the urine do not enable us always to locate the disease anatomically; in other words, we must ascertain further whether the disease involves the kidney, the bladder, or the lowest portion of the urinary tract. Such a decision is often very difficult, and occasionally is impossible. Renal tuberculosis, so far as its symptoms are concerned, often resembles almost to perfection vesical tuberculosis, and many patients have been cut for stone in the bladder when the result of the operation or the post-mortem examination revealed tuberculosis of the bladder. If the bladder-symptoms have been preceded by pain in the region of the kidney, if one or both of the kidneys are enlarged, or if an abscess has been formed, we have conditions upon which to base a diagnosis of a primary renal affection. Fragments of fibrin or caseous material in which renal epithelia are found point in the same direction. The amount of albumin in the urine must also be taken into consideration in making a diagnosis between renal and vesical tuberculosis. Nitze's cystoscope is a useful diagnostic instrument in the hands of experts. Finally, it may be stated that in all chronic inflammatory affections of the urinary organs it is necessary to make careful and often repeated examinations, both of the general and local symptoms, for the purpose of locating the disease as well as to determine its nature, which often can be done only by making a microscopical and bacteriological examination of the urine; if this should still leave the diagnosis doubtful, a resort to inoculation-experiments upon animals susceptible to tuberculosis becomes necessary as a decisive diagnostic test.

Prognosis.—The prognosis in vesical tuberculosis is always grave. Spontaneous recovery is exceedingly rare. As in tuberculosis of other organs, such a favorable termination is more likely to occur in children than in adults. Sir James Paget has met with two children in whom complete recovery without operative treatment took place. A case of partial recovery in an adult has been recorded by Thomas Smith. The patient was a man aged twenty-nine years, and was under treatment in St. Bartholomew's Hospital for symptoms resembling those of stone in the bladder. The diagnosis of vesical tuberculosis was strengthened by the existence of tubercular abscesses over the sternum and in the left upper arm. His other symptoms were cough, swelled testicles, frequent micturition, arrest of stream, hematuria, and pain in the left loin. The hematuria, contrary to the general rule, was aggravated by movements, and the pain was increased by emptying the bladder. No calculus could be discovered on repeated instrumental examination of the bladder. The symptoms were greatly relieved by palliative treatment, consisting in the administration of iron, cod-liver oil, and opiates. The hematuria yielded to confection of black pepper.

If the disease is mild and is limited to the bladder, it may remain in a latent condition for years. It is in such cases that the disease is most liable to be diagnosticated as chronic catarrh of the bladder, and the treatment adopted based upon such an erroneous diagnosis. In the more grave forms of the disease, usually complicated by tuberculosis in other organs, the symptoms are severe and follow in rapid succession when the disease is progressive, characterized by speedy local extension, general infection, and an early fatal termination. If the kidneys are affected at the same time, the course of the affection is greatly hastened, as in such cases general miliary tuberculosis is more likely to develop. Mixed infection, so often caused by careless and repeated examinations in search of stone in the bladder, aggravates the general and local symptoms and hastens the fatal termination. Suppurative cystitis complicating vesical tuberculosis is instrumental in hastening the extension of the tubercular process to the kidney, and is soon followed by a suppurative pyelonephritis.

There can be but little doubt that appropriate general and local treatment will prolong life and alleviate the distressing symptoms. The cases in which operative treatment has resulted in material palliation of symptoms, in prolonging life, or in effecting a permanent cure are so few and so far apart that the prognosis is not influenced to any considerable extent by it.

Treatment.—As has been stated before, the surgical treatment of tuberculosis of the bladder has so far not proved very satisfactory. The reasons for this are many. In the first place, the bladder performs an important function as a reservoir for the urine, and cannot be sacrificed without subjecting the patient to great discomfort, even should a radical operation for primary tuberculosis of this organ prove curative. The results of the experiments of

Tizzoni to make a new bladder out of a section of intestine are not sufficiently encouraging to make them applicable to man, in case the surgeon should decide to remove the bladder for malignant disease or extensive tuberculosis. The formation of a double external abdominal urinary fistula after extirpation of the bladder for otherwise incurable affections, even if the patient should recover from the operation, would give rise to so much discomfort to the patient, and cause him to be an object of such disgust to those around him, that even a permanent cure of the tubercular affection would be but an imperfect recompense for the loss of the bladder. Implantation of the ureters into the rectum or colon after complete extirpation of the bladder will probably soon be made practical by further perfection of the technique of the operation. The feasibility of the operation has been demonstrated by the experimental work of Tuffier and Reed, and a number of successful unilateral implantations of the ureter have been performed on man. The large intestine is the best substitute for the bladder. Partial extirpation of the bladder for localized primary tubercular affections of this organ would be an ideal curative operation, but such cases are rare, and a sufficiently early and positive diagnosis is difficult if not impossible.

An additional difficulty in the radical treatment of primary tuberculosis of the bladder by partial excision is the location of the disease, which usually involves the base of the organ—the part least accessible to radical treatment of this kind.

The greatest difficulty in the way of radical curative operations for vesical tuberculosis is, however, the fact that the disease is complicated from the beginning, or becomes so before a positive diagnosis is made, by tuberculosis of other organs. In view of the discouraging aspects of radical measures in the treatment of tuberculosis of the bladder, the general and local treatment increases in importance, and should be employed thoroughly and systematically from the very beginning of the disease, with the hope that the tubercular process may be arrested in its onward course and for the purpose of aiding nature's resources in effecting an occasional spontaneous cure. There is no organ in the body susceptible to tubercular disease in which, under favorable conditions, a spontaneous cure may not occasionally be effected. The local conditions surrounding the tubercular area in vesical tuberculosis are not favorable for such a termination without efficient medical treatment or early operative intervention.

The general treatment is of as great, if not greater, importance in the treatment of vesical tuberculosis as in pulmonary tuberculosis. The internal use of remedies known to exert at least an inhibitory action upon the growth of tubercle bacilli, such as guaiacol and creosote, should be commenced early and continued for a long time. The writer has seen most marked benefit from the uninterrupted prolonged use of guaiacol or its carbonate in the treatment of vesical tuberculosis. The dose should be gradually increased until 20 drops of the former or the same number of grains of the latter are reached, when the dose should be gradually diminished; but at no time should the use of the remedy be entirely suspended. If the appetite and digestion are diminished, the use of bitter tonics is indicated. Constipation always aggravates the vesical distress, and should be overcome by mild laxatives, such as Seidlitz powder, compound licorice powder, citrate of magnesia, etc. Anemia calls for some preparation of iron, as the tincture of the chloride of iron, ammonio-citrate, or tartrate of iron. Quinine in tonic doses is to be given if the patient is much debilitated, and in larger doses when the tubercular disease is complicated by suppurative cystitis. During the early stages of the disease the pain can be controlled to a certain extent by opium-and-belladonna rectal suppositories administered at bed-time, and, if necessary, in the morning, after the bowels have been moved spontaneously or by the aid of a rectal enema. A warm sitz-bath will often prove efficient in moderating pain and in diminishing strangury. A change of climate and the liberal use of slightly alkaline waters will prove useful in nearly all cases during the early stages of the disease. A diet composed largely of milk, cream, fresh meat, and fruit is indicated in all cases. Cod-liver oil appears to have been useful in some cases in diminishing the marasmus. The vesical irritation during the beginning of the disease is often relieved by 10-grain doses of boracic acid largely diluted, given four times a day; later, salol in similar doses proves more useful. Counter-irritation above the pubes results in more harm than good, while much benefit is often derived by the application of a compress wrung out of hot water. Bicycle- and horseback-riding are positively contraindicated, while moderate out-door exercise should be encouraged. Infusions of uva ursi, buchu, and triticum repens in 4-ounce doses often afford great relief when urination is frequent and painful.

Direct Medication.—Injections of solutions through a catheter have not proved as useful as parenchymatous injections in tuberculosis in other organs, for the reason that the tubercular cystitis is so often complicated by mixed pyogenic infection. In fact, in many cases the occurrence of suppurative cystitis could be traced to this method of treatment, the pus microbes having been introduced into the bladder by means of catheters or solutions imperfectly sterilized. Catheterization of the bladder in suspected cases of tuberculosis must be done with instruments that have been rendered absolutely sterile. The meatus should be well cleansed with an antiseptic solution before insertion of the instrument, and all solutions to be injected should be thoroughly sterilized. Incalculable damage has been inflicted by the use of contaminated instruments and solutions. This part of the treatment should be conducted by the attending physician or by a competent nurse, otherwise it will usually result in more harm than good. If the patient is so situated that he cannot avail himself of such services, he should be thoroughly instructed in the preparation and use of the instruments and the solutions to be used. If it is the intention to medicate the bladder through the urethra, a Nélaton catheter of large caliber should be used. Over-distention of the inflamed bladder always aggravates the local difficulties, and should be scrupulously avoided. Not more than an ounce or an ounce and a half should be injected at one time. The fluid should be allowed to remain for a few minutes, after which it may be allowed to escape. This procedure is repeated three or four times. The injection of air must be carefully avoided. The best instrument with which to make the injection is Thompson's rubber bulb supplied with a rubber or metallic conical point and stop-cock. This bulb as ordinarily made holds 6 ounces, and this quantity of the solution should be injected in four equal amounts. After removing the catheter it should be cleansed and kept suspended in a bottle containing a 5 per cent. solution of carbolic acid. Immediately before using it again the instrument should be rinsed in sterilized water and lubricated with carbolated lanolin or vaselin. The injection should be made once or twice a day.

Desnos[1] calls attention to the efficacy of corrosive-sublimate solution as advocated and practised by Guyon in the treatment of vesical tuberculosis. At first the injections are made weak— about 1 : 5000. If these are well borne, the strength of the solution is gradually increased to 1 : 1000. From two to five days is the time recommended between the applications. The writer has found this solution in the strength of 1 : 2000 of considerable value in favorably modifying the symptoms of the disease. . The tolerance of the bladder to the use of this solution gradually increases. It is advisable to commence the treatment with a solution

---

[1] *La Médicale moderne*, Jan. 28, 1894.

of 1:5000, used daily, gradually increasing the strength until a solution of 1:2000 is reached, increasing at the same time the interval between the injections from two to four days. It is good practice to follow the use of this solution by washing out the bladder with a physiological solution of salt. Nitrate of silver, so useful in the treatment of chronic vesical catarrh, has proved of no value in the treatment of tubercular cystitis.

Colin[1] has used with decided benefit injections of guaiacol. He finds that the injection of guaiacol carbonate has a very marked influence upon the pain, the frequency of micturition, and the state of the urine in all forms of chronic cystitis, but particularly in the tubercular variety. He employs a 20 per cent. solution in olive oil, 1 to 2 grams of this solution being injected once or twice daily. He also recommends the addition of iodoform (1 per cent.) as increasing the efficiency of the treatment.

The intravesical use of boracic, carbolic, and salicylic acids, creosol, creosote, lysol, and other antipyogenic agents is of use in the treatment of vesical tuberculosis complicated by secondary infection. Injections for such a purpose should be made at least twice daily, and if they should succeed in eliminating the cause and effects of the secondary infection, they prepare the way for a more successful employment of antibacillary injections.

Iodoform injections have probably been more frequently employed in the treatment of vesical tuberculosis than any other preparation since the antibac-

---

[1] *Journ. de Méd.*, Jan. 26, 1896.

illary effects of this remedy have been made known by Mosetig von Moorhof and Bruns. Emulsions varying in strength from 5 to 10 per cent. in oil and glycerin have been used for this purpose. Scientifically, this remedy is clearly indicated in the local treatment of vesical tuberculosis; practically, its use has been disappointing. Its most beneficial effects are observed in primary tuberculosis of the bladder uncomplicated by secondary infection. If this remedy is used as an intravesical application, the emulsion should be left in the bladder to be evacuated with the next passage of urine. This topical application should receive a fair trial in all cases in which operative treatment is contraindicated and when the tubercular process has not become complicated by pyogenic infection. A 10 per cent. iodoform-glycerin emulsion should be employed, and at least an ounce injected daily.

Another very valuable antiseptic for intravesical injection is the trichloride of iodin in the strength of $1/5$ to $1/2$ of 1 per cent.

Perineal Cystotomy.—Incision and drainage of the bladder through the perineum for vesical tuberculosis have been frequently performed, but the operation has seldom yielded the expected relief. The satisfactory results that follow this treatment in catarrhal and suppurative cystitis are in strong contrast with what is accomplished in tubercular disease. Perineal section has frequently been performed for an assumed suppurative cystitis, and the negative results of the operation have often led surgeons to make more careful inquiries into the nature of the affection, which inquiries not infrequently led to a correct diagnosis. The writer has performed perineal cystotomy in a number of cases of vesical tuberculosis with but temporary or no relief. Perineal drainage often requires long rest in bed, which is always a detriment to tubercular patients. The escape of urine on the sides of the drain into the bedding or clothing of the patient is an evil that cannot always be avoided, and a source of great discomfort to the patient. Washing out of the bladder can be done as efficiently through the urethra as through a perineal incision, and curettage by this route is dangerous and unsatisfactory.

Greiffenhagen[1] reports a case of bladder-tuberculosis permanently cured by perineal drainage and injections of a 5 per cent. iodoform-glycerin emulsion. The patient was forty-seven years old, married since he was nineteen years of age, and a farmer, by occupation. No hereditary tendency to tuberculosis existed in the family. Ten years ago he received a blow against the right side of the scrotum, followed by the formation of a painful swelling which after a while diminished in size somewhat; later it again became larger, and remained painful for nine years. Nine weeks before the operation he began to suffer from frequent micturition, the urine being stained with blood, and at the same time he noticed an induration at the tip of the glans and root of the penis. The symptoms gradually increased in intensity, attended by rapid emaciation. When first seen the patient had incontinence of urine, and every attempt to evacuate the

---

[1] "Zur chirugischen Behandlung der Blasentuberculose." *Deutsche Zeitschr. f. Chir.*, B. xliii. Heft 3.

bladder caused agonizing pain. No tuberculosis existed in any other organ. At the fundus and left side of the bladder a swelling as large as a walnut could be outlined by palpation. A smaller but similar swelling could be felt to the right and in front of the bladder. The mucous membrane of the urethra was swollen and inflamed, but there was no urethral discharge. The prostate gland was slightly enlarged and painful to touch. The scrotal swelling proved to be an ordinary hydrocele. The urine was turbid, neutral in reaction, and contained a trace of albumin, blood, and numerous tubercle bacilli. Median perineal section with drainage of the bladder was performed. For eight days the bladder was washed out through the drainage-tube twice a day. After this time the second irrigation was followed by the injection of 30–40 cubic centimeters of a 5 per cent. iodoform-glycerin emulsion. In the course of a month the urethra was dilated, causing a discharge in which, however, neither gonococci nor tubercle bacilli could be found. As soon as a No. 21 catheter could be inserted through the urethra the drain was removed, the catheter was permanently retained in the urethra, and the injections continued as before. Five weeks after the operation no tubercle bacilli could be found in the urine. In the course of a few months the fistula healed. The patient recovered completely, and remained in good health at the time the report was made. This was evidently a case of isolated tuberculosis of the bladder which yielded to incision and drainage combined with intravesical medication. It is doubtful whether the same result could not have been secured by urethral drainage and injections of iodoform emulsion through the catheter.

Fillipowo[1] recommends in tubercular cystitis injections of iodoform-glycerin emulsion, which in his practice proved of some benefit. Iodoformization has been practised more frequently through a suprapubic opening than through the urethra.

Suprapubic Cystotomy.—Incision and drainage of the bladder above the pubes will give more prompt relief and enable the surgeon to deal more efficiently with the tubercular lesion than cystotomy

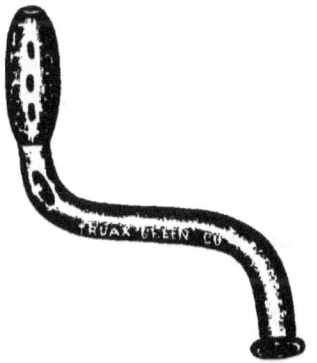

FIG. 18.—Senn's sigmoid self-retaining catheter for suprapubic drainage.

through the perineal route. The strangury which so often attends vesical tuberculosis when the disease has reached the neck of the bladder is usually relieved by suprapubic drainage. In a few cases a suprapubic fistula will not only relieve the patient of the intense vesical distress, but will prove a curative measure by securing rest for the diseased organ; but such cases are exceptions to the

[1] *St. Petersburger med. Woch.*, 1895, p. 310.

general rule. Guyon opened the bladder in cases of tuberculosis for the first time in 1885. In 1889 he made a full report of four cases treated by incision and drainage.[1] The first patient, a man twenty-four years of age, remained well four years after the operation. The second patient lived for two and a half years, when he succumbed to advanced renal tuberculosis. The bladder at the post-mortem examination showed a slight degree of inflammation of a non-tubercular character, while the remaining part of the genital tract was implicated in the tubercular process. In the third case the vesical disease had existed for nine months. The patient survived the operation for one year, when hectic symptoms developed without return of the bladder-symptoms. The suprapubic fistula remained open. At the autopsy the bladder-wall presented interstitial caseous foci; the mucosa was healthy; the kidneys were tubercular to a high degree. The fourth patient was thirty-four years of age, and was treated by perineal section, which he survived for three months. At the post-mortem the posterior part of the urethra and the base of the bladder were found thickly studded with tubercles. In this case the bladder was only incised and drained, no effort being made to remove the tubercular tissue or to apply topical remedies, while in the other cases curettage and local medication through the suprapubic opening formed part of the treatment.

Curettage and iodoformization are important aids

[1] "Résultats éloignés de quatre opérations pratiquées pour tuberculose vésicale." *Ann. des Mal. des Org. Gén.-urin.*, 1889, p. 642.

in the treatment of vesical tuberculosis after suprapubic cystotomy. The removal of tubercular tissue with the sharp spoon, followed by the local application of iodoform to the abraded surface, is well calculated to initiate a process of repair and to introduce an active phagocytosis. Through a large suprapubic incision the tubercular areas can be thoroughly exposed and the infected tissues subjected to mechanical removal with the sharp spoon. In cases where the tubercular process has penetrated deeply into the bladder-wall some care is necessary not to penetrate its entire thickness, as such an accident would be liable to be followed by tubercular or suppurative peritonitis.

J. L. Reverdin[1] calls the attention of the profession to the value of treating vesical tuberculosis through a suprapubic incision. The case upon which the paper is based was operated on more than two years before the date of the article, and was a case of secondary tubercular disease of the bladder. The treatment was by suprapubic incision, curettage, and cauterization. Both testicles and cords were affected. The prostate was involved. A very large perinephric abscess made its appearance later. There was considerable difficulty experienced in closing the suprapubic fistula, yet the patient at the end of two years retains his urine from one and a half to two hours and voids it without pain, the urine being clear and free from blood. Reverdin cites all the other unpublished cases to which he has had access, communicated by Roux,

---

[1] *Ann. des Mal. des Org. Gén.-urin.*, May-June, 1890.

making a total of 11 observations—Iverson, 1; Schutz, 1; Trendelenburg, 3; Guyon, 3.

Desnos,[1] in an advanced case of tuberculosis of the bladder, made a suprapubic incision and tamponed the bladder with iodoform gauze. The fistula remained open for a year, and was then closed by operation after the disease had disappeared. The patient recovered perfect health.

Verhaagen[2] subjected 3 cases of vesical tuberculosis to curettage. One case was materially benefited, while the improvement in the remaining cases was only of short duration.

Bangs[3] recommends as a palliative operation thorough drainage of the bladder, and in appropriate cases curettage through the suprapubic and perineal fistulæ. König succeeded in curing a case by curetting through a perineal fistula and using an iodoform-gauze tampon.

Ogier[4] reports 5 cases of tuberculosis of the bladder, from the clinic of Guyon, treated by the formation of a suprapubic fistula. The vesical wound was sutured to the skin, and the opening kept patent by the introduction through the fistulous opening, two or three times a day, of an ivory or metallic cylinder with a blunt point. The operation was performed for the relief of pain, and proved successful in most instances.

Bell[5] treated 3 cases of advanced tuberculosis of

---

[1] *Centralbl. f. Chir.*, 1893, p. 368.   [2] *Ibid.*, 1894, p. 44.
[3] *Ibid.*, 1893, p. 366.
[4] "Traitement de la cystite tuberculeuse par la cystostomie sus-pubienne. Création d' une urèthre contre nature." *Thèse de Lyon*, 1892.
[5] "The Treatment of Tuberculosis of the Bladder through a Suprapubic Section." *Journal of Cutaneous and Genito-urinary Diseases*, 1892, p. 292.

the bladder by suprapubic incision and the vigorous application of the thermo-cautery to the ulcerated surfaces. In 1 case the improvement that followed the operation continued for one and one-half years, and in the other cases material improvement could be ascertained three months after the operation. The bladder-symptoms never disappeared completely, but material improvement always rewarded the operative procedure. In 2 cases the wound healed completely without being followed by aggravation of the symptoms. The extension of the disease to the genital organs was in no instance retarded by the operation.

Guiard[1] recommends local treatment of the tubercular bladder through a suprapubic incision in cases in which the disease is limited to this organ. This method of treatment has also been carried into effect by Guyon, Reverdin, and Segond with varying results.

The surgical treatment of tuberculosis of the bladder in the female is the same as in the male, and in cases in which drainage is indicated the suprapubic route is much to be preferred, for various reasons, to the formation of a vesicovaginal fistula.

[1] Traitement chirurgicale de la tuberculose vésicale." *Ann. des Mal. des Org. Gén.-urin.*, 1888, p. 430.

PART X.

TUBERCULOSIS OF THE URETER.

PRIMARY tubercular ureteritis must be extremely rare; but the ureter, being the connecting-link between the kidney and the bladder, is frequently the seat of tuberculosis secondary to that of either of the organs, completing the tubercular area of the upper and middle portions of the urinary tract. If the tubercular process extends from the bladder in an upward direction, we speak of an ascending tubercular ureteritis; if the process starts in the pelvis of the kidney, and by continuity of surface finally reaches the bladder, a secondary cystitis follows a descending ureteritis. In bilateral tuberculosis of the kidney descending ureteritis usually occurs on both sides simultaneously or in more or less rapid succession (see Fig. 19). In primary tuberculosis of the lower portion of the urinary tract the disease may extend from the bladder along both ureters and reach the kidneys at the same time or in succession. A unilateral renal tuberculosis may reach the bladder by a descending tubercular ureteritis, and after infection of the bladder has occurred the process may ascend along the opposite ureter and involve the kidney on the opposite side. This is more likely to occur if the opposite ureter is the seat of a non-tubercular ureteritis resulting from a mixed infection of the bladder. That the ureteritis on the opposite side of a tubercular kidney and ureter in connection with a vesical tuberculosis may not be of a tubercular nature is shown by a case reported

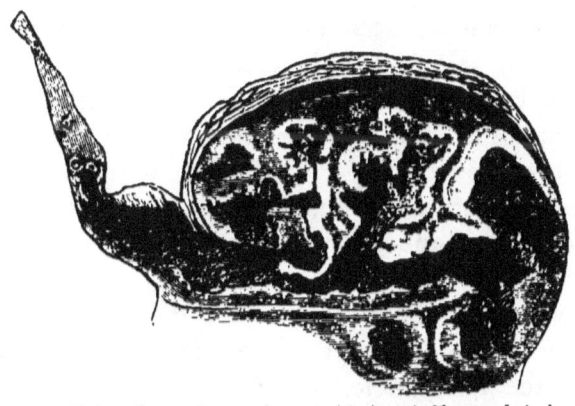

FIG. 19.—Tubercular nephro-pyelo-ureteritis (one-half natural size). The mucous membrane of the pelvis and the calyces is irregularly ulcerated; the dilated deep calyces are surrounded by a cheesy infiltration of the kidney-tissue that appears as a white border around them. In the ureter, near its pelvic dilatation, a few small round tubercular ulcers may be seen (Perls).

by Halle.[1] The post-mortem specimens were exhibited at a meeting of the Pathological Society of Paris. The right kidney was completely caseated; the corresponding ureter was the seat of an advanced tubercular infiltration and was reduced to an impervious cord; the bladder was the seat of great interstitial thickening, with tubercular deposits at its base, in the posterior part of the urethra, pros-

[1] *Bull. de la Soc. Anat.*, 1889, No. 26.

tate, and seminal vesicles. The opposite ureter and kidney showed marked suppurative changes, but no tubercular deposits were to be noticed in the structure of the ureter, and the character of the inflammation indicated a non-tubercular variety. Halle regards the events as of the following relation: A primary right renal tuberculosis; tubercular, obliterating, descending ureteritis; general cystic tubercular involvement; ascending, non-tubercular, left-sided ureteritis; and a non-tubercular left pyelonephritis.

This case illustrates well the familiar pathological fact that the type of the secondary pyelonephritis will correspond in perfection to the type of the ascending ureteritis that produced it. In a descending non-tubercular ureteritis the secondary affection of the bladder will correspond in character with the primary disease of the pelvis of the kidney that produced the ureteritis. It is not beyond the imaginable, in cases in which tubercle bacilli pass through the filter of a normal kidney and are brought in contact with an existing non-tubercular lesion of the ureter, that a primary tubercular lesion might become established in the soil prepared for the receptive growth and reproduction of the bacilli. In such a rare event the tubercular process would, in all probability, extend in both directions, ultimately involving the kidney on one side and the bladder on the other. Hydronephrosis of the kidney on the side opposite to a tubercular, renal, and ureteral tuberculosis may develop in the course of an ascending non-tubercular ureteritis. Adler[1] records a case in which advanced tubercu-

---

[1] *Medical Record*, July 7, 1889.

losis of the right kidney was associated with a hydronephrosis of the other, without tubercular manifestations noted elsewhere in the case. The ureter belonging to the hydronephrotic kidney must, therefore, have been the seat of a non-tubercular inflammation. Such a complication is, of course, possible only in case the bladder is infected with pus microbes, with or without an antecedent tubercular inflammation. Tuberculosis of the ureter appears usually as a diffuse affection in which the inner surface of the ureter is transformed into a superficial degenerated caseous mass that contains tubercles only in its outer portions. The walls of the ureter are greatly thickened, while its lumen is narrowed by the infiltration and the deposition of caseous material upon its surface. If the disease is very chronic, and the caseous deposits are washed away by the urinary current, and the exposed granulations are capable of conversion into scar-tissue, the lumen may become completely obliterated by cicatricial contraction. If the infection reaches the outer surface of the ureter, peri-ureteral foci complicate the case and become a serious source of tubercular infection of the tissues outside of the urinary tract. The lumen of the ureter in descending ureteritis may become suddenly and completely blocked by the infarction of caseous masses from the pelvis of the kidney, large enough to prevent their passage into the bladder. In instances of this kind the complete obstruction would be followed by retention of urine or of tubercular material above the seat of obstruction—an occurrence that would be indicated clinically by renal colic and the appearance of a swelling in the region of the kidney on the corresponding side.

Symptoms of obstruction to the escape of urine and tubercular cheesy material characterize both the ascending and descending forms of ureteral tuberculosis. In ascending tuberculosis symptoms of obstruction appear before the pelvis of the kidney is reached. Retention of urine gives rise to hydronephrosis. The swelling increases steadily in size in cases in which the obliterative process is progressive; it is intermittent if, from time to time, the lumen of the ureter is increased by the detachment and removal of the tubercular membranes that cause the obstruction. Descending ureteritis causes retention of urine and of caseous material. In advanced cases, in which the secreting structure of the kidney is completely destroyed, the kidney and the pelvis constitute a large sac filled with liquefied cheesy material, constituting an advanced stage of what is known as tubercular pyonephrosis.

The surgical treatment of isolated tubercular ureteritis will in all probability never be brought to the attention of surgeons, as, in the extremely rare cases in which the tubercular process has its primary seat in the ureter, the progressive infective process will have reached the bladder or the kidney, or both, long before a positive diagnosis can be made, and before the patient seeks surgical advice. In unilateral renal tuberculosis excision of the ureter should be done at the time the kidney is removed. This can be done most thoroughly through König's oblique incision. In women it is possible to remove the entire ureter by excising its lower end through a vaginal incision—a procedure that has its indications if the disease has not involved the bladder to any extent. In cases in which the disease has greatly debilitated the patient it might be advisable to remove the tubercular kidney and as much of the ureter as is within the reach of König's incision, and postpone for the necessary length of time the removal of the vesical portion through a vaginal incision. The use of the cystoscope is indispensable in deciding the propriety of removing the entire ureter in the operative treatment of unilateral renal tuberculosis. In cases of ascending tubercular ureteritis operative treatment is usually out of the question, as the disease of the bladder is, as a rule, sufficiently diffuse to involve both ureters simultaneously or in rapid succession.

In women the ureteral catheter will prove valuable in determining the location and extent of the lesion. By the employment of this diagnostic aid the surgeon is enabled to plan beforehand the nature and extent of the operation upon adjacent organs the primary seat of the tubercular process.

PART XI.

TUBERCULOSIS OF THE KIDNEY.

The etiology and pathology of tuberculosis of the kidney are now fairly well understood, and the frequency with which the affection occurs is being recognized by the general practitioner. The bacteriological examination of the urine, catheterization of the ureters as taught and practised by Simon, Pawlik, and Kelly, and the frequent employment of the cystoscope are modern diagnostic resources that have done so much in clearing up many doubtful points in the clinical phenomena of renal tuberculosis that the careful practitioner is enabled to make a positive diagnosis in most of the cases during the early stages of the disease. Before these methods of investigation were known and intelligently applied, when the physician had to rely on bedside observations, the differential diagnosis between renal calculus and tumor, catarrhal pyelitis and suppurative pyelonephritis, tumor, stone, and inflammatory affections of the bladder, and tuberculosis of the kidney was anything but satisfactory. The true nature of the disease was often suspected when associated with pulmonary phthisis, but the probable diagnosis could not be supported and confirmed by such reliable and conclusive means of examination. Patients suffering from renal tuberculosis were often repeatedly subjected to examination for stone in the bladder, to their great detriment, and a correct diagnosis was more frequently made in the post-mortem room than *intra vitam*. To Gustav Simon belongs the credit of having laid the foundation for more accurate means of diagnosis in the study and differentiation of renal affections, and his name will always be perpetuated in the history of surgery as the founder of rational renal surgery.

The relative frequency of renal tuberculosis as a primary disease, and the etiological relation of pulmonary and genital tuberculosis to secondary tuberculosis of the kidney, are subjects which are now attracting the attention of the profession, and which as yet are by no means definitely settled. Clinical experience and the results of post-mortem examinations are greatly at variance. Some authors claim that tuberculosis of the urinary organs seldom occurs as an isolated affection, while others, equally competent and reliable, maintain that the kidney is often the seat of primary tuberculosis, and that the process extends from here to the lower portions of the urinary tract and the male genital organs. Another class of clinicians and pathologists hold that in the male the genital organs are most frequently affected primarily, and that the urinary tract is involved secondarily by an ascending inflammatory process. Evidences and statistics on both sides will be quoted, when it will become apparent that neither of these conclusive positions can be maintained by reliable clinical statistics nor the results of post-mortem examinations; facts will show that renal tuberculosis frequently gives rise to secondary tuberculosis of the male genital organs, and that in primary tuberculosis of the male genitals the tubercular process often ascends and involves successively and by continuity of surface the bladder, ureters, and finally the kidneys. Collinet[1] made a careful literary investigation of the origin and dissemination of genito-urinary tuberculosis, representing the views mainly of French and English surgeons. The following are some of the important points which he brought out: The tubercular process may commence in any part of the urogenital tract independently of pulmonary tuberculosis. One out of 18 consumptives suffers from some form of genito-urinary tuberculosis. The kidney is affected either primarily or simultaneously with other organs —namely, the ureters, bladder, prostate, and least frequently the testicles. The ureters are never attacked primarily, but always in connection with renal tuberculosis. The testicle is often the starting-point of tuberculosis of adjacent organs, especially the prostate, less frequently the kidney. The vas deferens is seldom the seat of primary tuberculosis; the process extends to it at either of its ends from adjacent connecting organs. The prostate and the seminal vesicles may be the seat of primary infection, but more frequently with simultaneous tuberculosis of the kidneys and testicles, in which case infection is not by continuity, but collateral. The

---

[1] "Considérations sur le tuberculose des organes génito-urinaires chez l'homme." *Thèse de Paris*, 1892.

bladder is seldom affected primarily, but usually consecutively to kidneys and prostate. The urethra is very rarely the seat of primary tuberculosis. With few exceptions, the sentiments of the German surgeons on this subject are advanced by Oppenheim,[1] who collected 60 cases of urogenital tuberculosis—37 males and 23 females—for the purpose of studying from a pathological standpoint the anatomical discrimination of the tubercular process. He admits that primary tuberculosis of the kidney can occur by elimination of tubercle bacilli through the kidney, when some of the bacilli become arrested in the capillary network of the parenchyma, causing miliary tuberculosis. The tubercular pyonephrosis —the form that interests the surgeon—he regards as a secondary affection, the result of an ascending tubercular process from the genital tract. He states that in such cases the bacilli, after they have left the kidney, produce a tubercular cystitis or prostatitis, in consequence of which retention of urine occurs, the retained fluid finally reaching the pelvis of the kidney. The result of pressure on the pelvis is a flattening of the papillæ of the kidney, which then constitute a *locus minoris resistentiæ* in which the bacilli begin their work of destruction.

Collmet, in 70 cases of urogenital tuberculosis, found affections of the kidney in 53 cases; of the testicles in 41; of the prostate in 44; of the kidneys, testicles, and prostate in 20; of the testicles and prostate in 6; of the kidneys and prostate in 16; of the testicles in 9; of the kidneys in 10; of

---

[1] "Zur Kenntniss der Urogenitaltuberculose." Dissertation, Goettingen, 1890.

the kidneys and testicle in 7; and of the prostate in 2 cases. In 59 cases of tuberculosis of the prostate the seminal vesicles were involved in 36. Of his 70 cases of urogenital tuberculosis, the bladder was affected 33 times; simultaneous renal tuberculosis existed 29 times; simultaneous prostate tuberculosis, 24 times; simultaneous testicle tuberculosis, 11 times; simultaneous affection of all these organs, 9 times.

The result of investigation will show that the kidney may become infected with tubercle bacilli— 1. From the blood, without demonstrable tuberculosis in any other part or organ of the body; 2. From the blood, as a secondary affection in pulmonary phthisis and tuberculosis of any other organ; 3. By continuity of surface in ascending tuberculosis of the male genital organs. It is equally well established that in the male primary and secondary tuberculosis of the kidney very frequently, if life is sufficiently prolonged, gives rise to genital tuberculosis.

Frequency.—The comparative frequency of tubercular disease of the kidney may be judged by the following statistics quoted by Roberts: "Embracing both primary and secondary deposits—the latter being, especially in children, by far the most frequent—out of 1317 tubercular subjects examined in the Pathological Institute of Prague (out of a total of 6000 bodies), tubercle in the kidney was found 74 times, or in the proportion of 5.6 per cent. of tubercular subjects.[1] Among 315 tubercular chil-

---

[1] *Prager Vierteljahrsschrift*, B. i. p. 1, 1856.

dren, Rilliet and Barthez found tubercle in the kidneys 49 times, or in the proportion of 15.7 per cent." The observations of Dickinson relative to frequency of renal tuberculosis in tubercular subjects correspond very closely with those of Rilliet and Barthez. Morris shows, by an examination of the records of the Middlesex Hospital, that of the 44 cases of tubercular kidney met with in 2610 necropsies, 29 were miliary and 15 were caseous. Of the 29 cases of miliary tuberculosis, 18 were males and 11 females, and in 28 of the 29 cases both kidneys were diseased; while in the 15 cases of primary renal tuberculosis (9 males and 6 females) both kidneys were diseased in 8 cases. More careful bacteriological and histological examinations of kidneys removed for disease and obtained by postmortem examinations will undoubtedly add materially to the comparative frequency of renal tuberculosis, as the real nature of the disease will be recognized in many specimens in which it was formerly overlooked or sought for with insufficient care.

The diagnostic aids that have been referred to have done much to enlarge the field of surgery of the kidney. The cystoscope is useful in inspecting the urine as it escapes from the ureteral orifices, and in demonstrating tubercular lesions of the bladder, which so often complicate primary renal tuberculosis. Catheterization of the ureters as practised in the female affords an opportunity to obtain the urine from each kidney separately for microscopic and chemical examination. In the male this method of examination has not been sufficiently developed for practical purposes.

Bacteriological examination of the urine is of the utmost importance in the early diagnosis of tuberculosis of the kidney. Koch was the first to succeed in cultivating tubercle bacilli from cases of tubercular pyelitis. Although tubercle bacilli may be found in the urine of patients suffering from tuberculosis of the lungs and other organs, and they are not easily cultivated from the urine of patients suffering from renal tuberculosis, as putrefactive bacteria may contaminate the culture and destroy the specific bacilli, their presence in the urine of persons suffering from symptoms of urogenital disease is very suggestive of tuberculosis of the kidney or bladder. Exclusive reliance cannot be placed on the positive results of this method of examination, as smegma bacilli have been mistaken for tubercle bacilli—an error which, of course, would lead to wrong diagnostic conclusions. Leyden and Mendelsohn relate the case of a woman suffering from renal calculus, but which the attending physician, deceived by the bacilli that he found, diagnosed as tuberculosis. In this case the value of catheterization of the ureters was shown most strikingly, for the seat of the disease was thereby exactly located, the healthy condition of the other kidney was demonstrated, and a successful operation for the removal of the stone was performed.

Trisch[1] has written a very instructive paper on the diagnostic value of bacteriological examination of the urine in cases of urogenital tuberculosis. The symptoms, exclusive of the presence of tubercle

---

[1] *Internationale klinische Rundschau*, July 12, 19, 26, 1892. Sajous, *Annual of the Universal Medical Sciences.*

bacilli in the urine, are of very variable character and occurrence, and, therefore, never can be thoroughly relied upon for their diagnostic significance. Even the recurrence of tubercle bacilli is not to be regarded as a positive sign, since it has been demonstrated that in tuberculosis elsewhere than in the urogenital organs the urine may contain bacilli, and that unless the urine is obtained directly from the kidney, smegma bacilli might be mistaken for tubercle bacilli. The author states that the discovery of bacilli in the urine is apt to be attended with difficulty both from the scarcity and the enormous numbers to be seen in different specimens at different examinations. A special grouping of the bacilli—that in which they are arranged in S-shaped groups—is especially valuable as a diagnostic sign of urinary tuberculosis (see Fig. 20); but it is probable that even in such groups the tubercle bacilli may find their way into the urine from more distant tubercular foci. Nevertheless, these groups are more apt to come from masses of bacteria in close contact with the urinary system.

In examining the urine for bacilli the author mentions the methods suggested by Kirstein, Biedert, Sehlen, Wendriner, and Philip. Some of these methods were originally suggested for the purpose of discovering the bacilli in sputum, but with slight modifications they have been found well suited to urinary examination. Where the urine contains large quantities of mucus the author suggests that it be treated with an alkali to dissolve the latter, thus permitting the precipitation of the bacilli. When the sediment is made up largely of urates and other crystalline substances he uses, after Wendriner, a solution composed of hot distilled water in which is dissolved 12 per cent. of powdered borax, and afterward an equal amount of boracic acid is added. The solution, added to the urine, dissolves the uric acid, urates, earthy phosphates, and the organized substances. The urine is permitted to stand until the sediment is deposited; then the supernatant fluid is poured off, the sediment is washed and collected upon a filter, and a portion is pressed between two covers, dried, and stained like sputum. The method of examining urine for tubercle bacilli is unquestionably the most reliable in its results, but as it is quite complicated, it will be resorted to only in the laboratory and in cases in which simpler procedures have yielded negative results. If the bacilli are scanty in the urine of suspected cases of renal tuberculosis, the employment of the centrifuge is a very great aid in demonstrating their presence.

If tubercle bacilli are found in the urine obtained from the bladder or kidney by catheterization under careful antiseptic precautions, positive proof has been furnished that the patient is tubercular, and it then remains for the physician to locate the disease. If the bacteriological examination proves negative, it does not necessarily show that the patient is not suffering from tuberculosis of the urogenital organs, and if the clinical symptoms point in this direction, inoculation-experiments with the urine will frequently yield results that must be regarded as final in the diagnosis between tuberculosis and other inflammatory affections of the genito-urinary system.

For the surgeon it is absolutely necessary, before he suggests operative interference of a radical nature in the treatment of renal tuberculosis, not only to ascertain the character of the disease, but also to satisfy himself, by different methods of examination that will be referred to further on, that the disease is limited to one organ, as a nephrectomy in a case in which both organs are affected would necessarily be followed by death from uremia.

Pathology.—From a clinical and pathological basis tuberculosis of the kidney should be classified into —1. Miliary tuberculosis; 2. Caseous nephritis. 3. Tubercular pyelonephritis. Some authors are satisfied to make a division into the acute miliary and the chronic or caseous variety, while others, like Tuffier,[1] make a very elaborate classification. He subdivides the chronic form into (*a*) pyelonephritis; (*b*) tubercular nephritis; (*c*) tubercular hydronephrosis; and (*d*) extensive tubercular cavities or diffuse caseous tuberculosis, the last two forms being the result of ascending tubercular ureteritis. He considers that changes in the ureters —consisting of ureteritis and its sequelæ and the obliteration of the ureter, partial or complete—are lesions very often overlooked, although their consequences are serious. In ureteritis there is pyuria; in obliteration or stenosis there is enlargement of the kidney. When the kidney is not structurally involved in tuberculosis, hydronephrosis results; but when the kidney is the seat of tuberculosis, the

---

[1] *Archives Générales de Médecine*, May, 1893.

organ is progressively invaded, and caseation and abscess-formation follow step by step. The classification suggested by Duret[1] is still more confusing. Besides the miliary form, he recognizes the cavernous form with multiple foci, the cavernous form with a single cavity or pouch, and the cavernous form with pyelonephritis. To these he added the complicating varieties, perinephric abscess, adiposclerosis of the fatty capsule, and fistula, or the fungous form. The classification of renal tuberculosis given above not only indicates the manner in which the infection has occurred, but also locates the part of the kidney first affected, and is especially useful from a surgical point of view. In miliary tuberculosis infection takes place through the blood, the disease is bilateral, and the patient succumbs in a short time to the effects of a diffuse miliary tuberculosis which involves many organs predisposed to localization of the tubercle bacillus. In caseous nephritis the disease is often limited to one organ, is chronic in its course, and infection has occurred either through the blood, with or without tuberculosis in any other organ, or from the pelvis of the kidney, through the lymphatics or connective-tissue spaces. In tubercular pyelitis the disease involves the surface of the pelvis of the kidney, is frequently unilateral, and infection is caused either by an ascending tubercular process from the lower urinary tract along the ureter, or by the precipitation of tubercle bacilli eliminated by the kidney upon the surface of the pelvis, at points where anatomical or

---

[1] *Journ. des Sci. Méd. de Lille*, June 24, 1894.

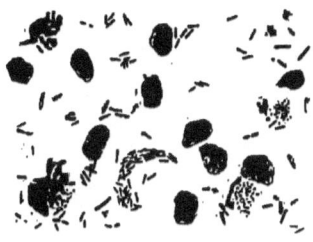

Fig. 20.—S-shaped groups of tubercle bacilli as seen in some cases of renal tuberculosis (von Jaksch).

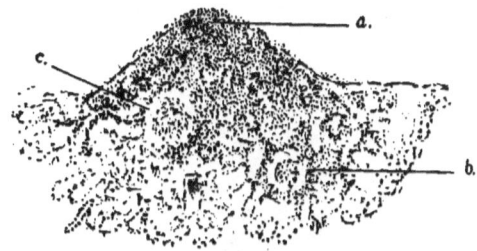

Fig. 21.—Miliary tubercle in kidney: *a*, area containing bacilli; *b*, shrunken glomerulus; *c*, glomerulus, coagulation-necrosis.

acquired conditions serve as determining causes for the localization of the microbes.

Miliary Tuberculosis.—In miliary tuberculosis of the kidney from blood-infection both organs are usually affected simultaneously, and the earliest inflammatory changes are observed close to and in the course of the smallest arterial vessels (see Fig. 21); the disease therefore appears anatomically as an interstitial process. So far, pathologists have failed to find the histological evidences of an early endovascular inflammation such as has been described by Hektoen in tubercular meningitis, and which undoubtedly often precedes the interstitial inflammation so constantly described by authors as the first tissue-change in renal tuberculosis. In miliary tuberculosis the bacilli are first brought into contact with the intima of the small arterial vessels by an embolic process or implantation upon the intima, and it is therefore reasonable to suppose that textural changes in the vessel-wall at the different points of infection should and do precede the interstitial formation of miliary tuberculosis. Virchow called special attention to the changes that occur in the connective tissue during the first stages of tubercular inflammation of the kidney. The description of miliary tuberculosis as given by Newman[1] shows clearly the connection of the tubercle-formation with the blood-vessels. He regards miliary tuberculosis as an infective process giving rise to a general eruption of tubercles in the various organs of the body. In such cases the tubercle

[1] *Lectures to Practitioners on the Surgical Diseases of the Kidneys.* London, 1888, p. 281.

bacilli are conveyed to the kidney by the blood-vessels, and hence the favorite seat of the disease is in the cortex, where the tubercles appear as small, semi-translucent, grayish streaks surrounded by a darker zone of deeply ingested renal tissue (see Fig. 22). These miliary nodules are frequently arranged in rows following the direction of the blood-vessels, and present an appearance closely resembling minute hemorrhagic infarcts. This regular arrangement is seen in a few cases, but in most of the cases that came under Newman's observation the tubercles were distributed without any evident relationship to the arterial blood-supply of the kidney, while in other instances the infective process appeared as if limited to an area supplied by a single branch of the renal artery. In this form of tuberculosis the disease is limited to the parenchyma of the kidney, and does not extend to the lower urinary tract.

It would be interesting to study from the earliest stage the minute pathological changes that occur in vessel-walls and the parenchyma of the kidney in acute miliary tuberculosis in animals in which the disease has been artificially produced. By the growth of the tubercular nodule the adjacent blood-vessels are pushed apart, the nutrition of the tissues is impaired, and the adjoining glandular tissue is likewise subjected to the harmful effects of pressure. The nodules are most numerous underneath the capsule, and are often found around the glomeruli. The cortex is more diffusely infiltrated than the pyramids. The histological structure of the tubercle tissue is the same as in other organs. In the formation of

FIG. 22.—Numerous subcapsular nodules from a case of miliary tuberculosis. The lower figure represents the kidney laid open, showing the nodules in the center.

the nodules the epithelia of the uriniferous tubules and the glomeruli take an active part, and from the epithelioid cells resulting from their tissue-proliferation giant cells are often produced. In the infected territories the blood-vessels, in consequence of pressure and the action of the toxins of the tubercle bacillus, become occluded by thrombi, and as a consequence the tissues undergo necrosis. The longer life is prolonged, the larger the nodules will grow by coalescence and tissue-proliferation in the immediate vicinity of the different foci of infection. Life is usually destroyed before well-marked caseation sets in. In acute miliary tuberculosis the kidney is not sufficiently enlarged by the inflammatory process to be distinctly felt by bimanual palpation.

Caseous Nephritis.—Caseous nephritis may occur as a primary isolated affection, with or without tuberculosis of any other organ, by infection through the blood or in consequence of an extension of the tubercular process from the pelvis of the kidney through the lymphatics. It is of importance and interest to the surgeon because it frequently affects only one kidney, and therefore under favorable conditions it is amenable to successful treatment by operative interference; and in cases in which it follows in the course of an ascending tubercular process from the genital organs or bladder, the early and effective treatment of the primary disease constitutes the best possible prophylactic measure. In this form of renal tuberculosis the infection occurs in the substance of the kidney, and from one or numerous foci the disease involves the surrounding tissues until, if life is sufficiently prolonged, the entire organ is destroyed and converted into a large tubercular abscess (see Fig. 23). If the disease has a parenchymatous origin, the pelvis of the kidney becomes infected as soon as it is invaded by the advancing tubercular process. Virchow[1] described this form of tuberculosis of the kidney years ago under the name of *nephro-phthisis*.

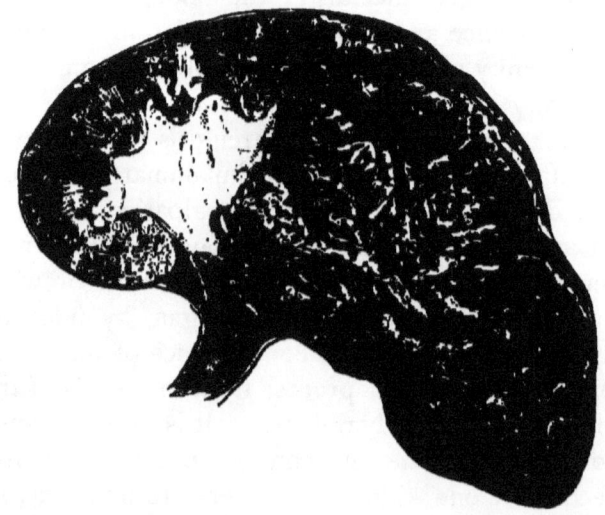

FIG. 23.—Chronic caseous nephritis and perinephritis which have resulted in the destruction of more than half of the organ; specimen obtained by operation in the clinic at Greifswald (Schmid).

He says that by careful study of many cases he found, in the kidney, tuberculosis in all its stages, from the gray miliary nodules to groups of nodules, and from groups to diffuse infiltration, until finally the whole organ was implicated and destroyed.

[1] *Die Krankhaften Geschwülste*, B. ii. p. 655.

What occurs in mucous membranes upon the surface takes place in the kidney in all directions, so that finally the melting mass empties itself into the calices of the pelvis, when at last a large sac filled with caseous material takes the place of the kidney. He insisted very strongly upon the fact that the tubercular process did not commence in the secreting structure of the organ, but in its stroma, the interstitial tissue, where nucleus and cell-formation are in progress, resulting in the production of the primitive nodules; the spaces between the uriniferous tubules and the glomeruli become wider and wider; the epithelial cells disintegrate into a pultaceous mass. The tunicæ propriæ, and finally the blood-vessels, are destroyed, and at last nothing is left but the proliferating new formation.

The older writers included this form of renal tuberculosis under the head of scrofula, and even as late as 1888, Iscovesco[1] describes scrofulous nephritis as a diffuse process that manifests itself to the naked eye as a large white kidney, the lesions being parenchymatous and interstitial. He also made the statement that clinically such a condition may remain latent indefinitely or may manifest itself promptly by albuminuria and the ordinary symptoms of the large white kidney.

What has been known and described as "scrofulous inflammation of the kidney" corresponds with what is now called "chronic caseous nephritis." The chronic form of renal tuberculosis, irrespective of its anatomical starting-point and manner of in-

[1] *Contribution à l'Étude de la Néphrite Scrofuleuse.* Paris, 1888.

vasion, often results in extensive caseation and the formation of large cavities in which sometimes numerous S-shaped groups of bacilli are found, re-

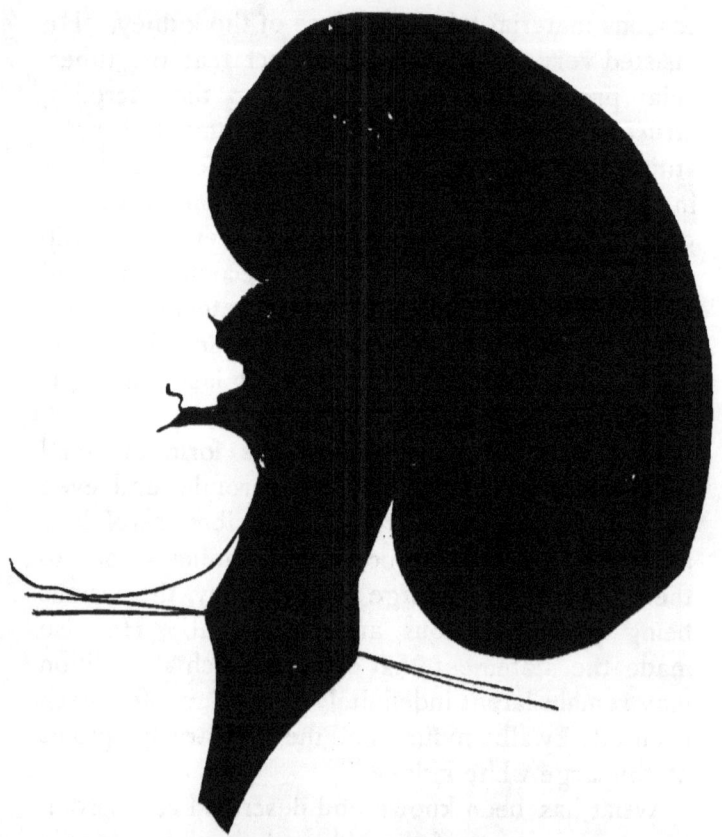

FIG. 24.—Rupture of a tubercular abscess of the spine into the pelvis of the kidney, causing tuberculosis of the kidney, pelvis, and upper portion of the ureter; specimen obtained post mortem (Schmid).

sembling in many respects the cavities in tubercular lungs; hence the terms *renal phthisis* and *nephrophthisis*. In primary chronic renal tuberculosis the tubercular process sooner or later reaches the pelvis of the kidney, causing tubercular pyelonephritis, which, if life is prolonged, descends by continuity of surface, involving in succession the ureter, the bladder, and in men the genital organs. Perforation of a tubercular abscess into the pelvis of the kidney during life is announced by the appearance in the urine of tubercular detritus and an increase in the number of bacilli (see Fig. 24).

The chronological order of infection of the different parts of the urinary system can usually be determined by the pathological conditions of the organs or specimen removed by operation, the part first involved presenting the most marked evidences of caseation and disintegration. In ascending tuberculosis in which the parenchyma of the kidney is reached last, the lower portions of the urinary tract show the most advanced stages of caseation and ulceration. If the kidney is primarily affected, swelling of the organ is moderate, and its detection by palpation may be impossible. The surface of the kidney becomes uneven and nodular, and the capsule, especially at the points corresponding with the peripheral cheesy masses, is not only decidedly thickened and at times of cartilaginous consistency, but also contains cheesy foci. Perforation of such peripheral foci not infrequently leads to tubercular perinephritis and the formation of a tubercular perinephric abscess. If the renal process assumes a very chronic type, with the thickening of the capsule a cirrhosis of the adipose capsule takes place by an active connective-tissue proliferation, the product of which unites with the outer surface of the capsule, so that the tubercular product is imprisoned in a mass of connective tissue which for the time being protects the surrounding tissues against infection. It is in such cases that removal of the kidney proves an exceedingly difficult task, and the operator is often obliged to perform subcapsular nephrectomy, as first suggested and practised by Ollier.

Tubercular Pyelonephritis.—Tubercular pyelonephritis and its ultimate resulting condition, tubercular pyonephrosis, are caused either by the escape of the contents of a primary tubercular focus in the kidney into the pelvis, primary infection of the pelvis, rupture into it of a tubercular abscess from an adjacent organ, or an ascending tubercular infection from the lower part of the urinary tract. The pathological conditions necessarily vary according to the manner in which the pelvis was invaded. In secondary pyelonephritis from a primary renal focus the ureter, as a rule, remains patent until the parenchyma of the kidney is extensively destroyed and the capsule of the kidney has undergone thickening, so that any considerable distention from the retention of urine cannot occur. The enlargement of the organ, which is seldom considerable, takes place by the production of tubercle-tissue in the kidney and pelvis and the formation of cheesy masses. If liquefaction of the caseous material takes place and the ureter remains patent, the liquefied tubercular material finds its way to the bladder and is discharged with the urine, infecting the mucous surfaces over which it passes. The kidney is converted into a pus-sac only in case the lumen of the ureter becomes sufficiently narrowed to prevent the escape of urine and the softened cheesy material. If the ureter is completely blocked by impaction with tubercular product, or if cicatricial stenosis of any of the secreting structures of the kidney takes place, secretion is soon arrested by pressure, and the contents of the sac are composed exclusively of the products of the tubercular inflammation. The contents of such a pyonephrotic sac are, of course, modified by a complicating secondary infection with pus microbes. It is in cases of this kind that the symptoms, local and general, as well as the pathological conditions, are typical of what is meant by the term tubercular pyonephrosis. It is likewise in such cases that a perinephric abscess is most likely to develop, complicating the diagnosis.

Primary tuberculosis of the pelvis of the kidney is rare as compared with the secondary form. It is caused by the localization, upon the surface of the pelvis, of tubercle bacilli that have passed the capillary filter of the kidney and have found their way into the uriniferous tubules and thence into the pelvis, or directly into the pelvis. This form of renal tuberculosis is as often bilateral as unilateral, and is a disease of adolescence and middle age rather than of childhood. The parts of the pelvis predisposed to localization are the papillæ. The process begins in the apices of the papillæ by the formation of isolated groups of tubercles, which are soon transformed into a caseous mass that softens and infiltrates the adjacent tissues. Extension to the parenchyma of the kidney also takes place through the lymphatics and blood-vessels. The caseous material often becomes the seat of calcareous deposits, creating a condition that may easily be mistaken for stone. In this class of cases fragments of broken-down tubercle-tissue and bacilli may be detected in the urine almost from the beginning of the disease. As soon as the disease involves the ureter, symptoms of obstruction set in, and retention of urine and the accumulation of caseous material result in more or less enlargement of the kidney. As the disease progresses renal foci coalesce, the parenchyma of the kidney is rapidly destroyed, and in advanced cases no trace of it remains, the kidney and pelvis being transformed into a common tubercular cavity. The tubercular process at an early stage descends, and often involves in a remarkably short time the ureter and the bladder. The tuberculosis of the kidney resulting from an ascending infection, the tubercular pyelitis that follows when the kidney is reached is preceded and complicated by pathological conditions of the ureter that antedate the renal affection. Distention of the pelvis of the kidney from the retention of urine caused by the ureteritis flattens the papillæ, thus preparing the way and hastening the extension of the infection to the pelvis of the kidney.

In men the primary starting-point of this form of renal tuberculosis is often the bladder or some part of the genital organs. Anderson[1] reports a number of such cases. The precedence of vesical symptoms in all these cases is apparently overlooked, although the post-mortem evidence clearly indicates considerable duration of vesical inflammation from the in

---

[1] *Glasgow Med. Journ.*, July, 1890.

tense thickening of the walls of the bladder. The fact that only one kidney shows evidences of tubercular invasion is by no means inconsistent with the upward advance of the process, and would be directly opposed to urinary infection through the general circulation. Why vesical tubercular inflammation should advance in some cases to the invasion of but a single ureter, pelvis, and kidney should not be regarded as difficult of explanation. Either the inflammatory process affects one ureter by accident, or because of less resistance of the tissue to passive ureteral dilatation following vesical spasm, or as a result of spasm of the bladder, the downward flow of urine is seriously interfered with and stagnation takes place, and the infection travels in the direction offering the least resistance.

In renal tuberculosis following in the course of an ascending tubercular ureteritis early distention, and later great distention, of the pelvis take place. The ureter is gradually transformed into a more or less rigid tube with thickened walls and narrowed lumen. The lumen of the tube is narrowed by tubercular deposits upon its surface; thickening of its walls and cicatricial contraction result in partial and sometimes in complete obstruction. To external appearances the ureter may appear, however, as if it were dilated, as in hydronephrosis, but on careful examination it is found that its walls are thickened and its lumen is narrowed. If the ureteric obstruction is sufficient in degree to interfere with the free flow of urine before the pelvis of the kidney is reached by the tubercular infection, a simple hydronephrosis is caused, which, when the retained urine becomes mixed with the tubercular product of the secondary tubercular pyelitis and nephritis, is transformed into a tubercular pyonephrosis. It is in this form of renal tuberculosis that the kidney attains often an immense size, in case the opposite organ is not affected and is capable of assuming the requisite compensatory function. The miliary tubercles soon spread over the entire surface of the pelvis, undergo speedy caseation, and by softening of the caseous material there are developed ulcers which become confluent and form large irregular surface defects. The infection of the kidney from the pelvis spreads through the lymphatics and the blood-vessels in very much the same way as septic inflammation in suppurative pyelitis. The miliary tubercles which develop in the course of blood-vessels and lymphatics are transformed into a caseous mass. In the course of a few weeks or months this mass forms an irregular softened area, which by progressive peripheral infiltration spreads inward and in the course of time destroys the entire parenchyma of the kidney, the pelvis and the kidney being converted into one common tubercular cavity with characteristic contents and miliary tubercular infiltration of its thick walls. When the kidney is converted in this manner into a sac by the tubercular process, the sac is incompletely divided by a few connective-tissue septa—the remaining renal calices —and forms a single compartment with the renal pelvis, which has undergone similar changes.

Owing to the unequal distribution of the tubercular process and the resistance offered to distention by the capsule, the shape of the kidney is often materially altered. The pouches that form correspond with the prominences upon its surface, so that the sac is very irregular in its outlines. In tubercular pyonephrosis with incomplete obstruction of the ureter the contents of the sac escape in part, at different times, with the urine, so that, in case the opposite organ is in a healthy condition, the urine may be clear for a number of days, when suddenly it becomes turbid for a time by the admixture of tubercular detritus from the affected kidney. The tension caused by retention is often the cause of renal colic, which is relieved by the periodical discharge of the contents of the sac. In a few cases the connective tissue surrounding the ureter is found infiltrated with pus, and occasionally, when there is a simultaneous affection of the bladder, the perivesical connective tissue is found in the same condition. In the latter case regional infection of the lymphatic glands is often met with.

In chronic tuberculosis of the kidney the opposite organ is not infrequently infected by way of the bladder and the opposite ureter, a descending tuberculosis being followed by an ascending process along the opposite ureter. In unilateral tuberculosis of the kidney the opposite organ increases in size and assumes compensatory function. The compensatory hypertrophy reaches the maximum degree in cases in which the renal affection is primary and very chronic. The hypertrophic kidney occasionally becomes the seat of a later affection. Camargo[1] describes a case of hypertrophy with tuberculosis

---

[1] *Revue Méd. de la Suisse Romande*, October, 1892.

of one kidney following atrophy of the other. The organ increased in volume, measured 17 centimeters in length, 7 centimeters in width, and 6.5 centimeters in thickness; the size was 310 cubic centimeters, while that of the normal kidney, according to Vierordt, is 149 cubic centimeters. The glomeruli were also increased in diameter, measuring on an average 320 millimeters, the normal average being 200 millimeters. The hyperplasic kidney was at the same time tubercular. Camargo defends the theory that vesical tuberculosis is almost always, if not always, descending—*i. e.* follows renal tuberculosis—and that, therefore, surgical interference can be attended by no lasting results.

Absence or destruction of one kidney is often attended by cardiac hypertrophy. In 4 cases of absence of one kidney collected by Simon, the heart was found hypertrophic in 2. In the 2 cases in which the heart was normal the single kidney was found greatly increased in size by compensatory hypertrophy—double the size of a normal kidney. The vicarious growth of the functionating kidney was the cause of absence of enlargement of the heart.

Etiology.—The essential causes of renal tuberculosis are susceptibility of the kidney to tubercular infection, and the presence in the organ of a quantity of tubercle bacilli of sufficient quantity and virulence to produce their specific pathological effect. Among the exciting causes must be enumerated trauma, congenital or acquired pathological conditions that determine localization of the tubercle bacilli, and anatomical abnormalities of the kidney.

Antecedent inflammatory affections (especially of the pelvis) and trauma are the conditions that most frequently determine tubercular infection. The influence of trauma in determining localization is well illustrated by a case reported by Faulds.[1] The patient was a man twenty-one years of age. The first symptoms of the disease manifested themselves a short time after an injury received while playing foot-ball, a companion falling upon and striking him in the left lumbar region with his knee. The prominent symptoms were pain on micturition, hematuria (occasionally), pyuria, loss of strength and flesh, and tenderness over the left kidney. Dr. J. D. Bryant made an exploratory incision, and found the left kidney enlarged and so firmly bound down by fibrous adhesions that he declined to remove the organ. Death followed in several months, and at the autopsy the organ was found in an advanced state of caseation. The opposite kidney contained a few tubercular nodules. The mesenteric glands were enlarged and cheesy. The spleen contained a few isolated nodules. At the apex of the left lung there was a small cavity. The writer regards the case as one in which the trauma acted as a potent cause in determining the renal disease.

Before considering the different avenues of infection, it is advisable to refer to some of the experimental researches that have been made to ascertain that tubercle bacilli are eliminated through the kidneys and pass off with the urine.

---

[1] *Lehigh Valley Med. Mag.*, April, 1892.

Rosenstein[1] and Babes[2] found tubercle bacilli in the urine of tubercular patients. Fardel[3] refers to the condition of the kidneys in cases of miliary tuberculosis as an evidence that the tubercle bacilli are diffused through the blood-vessels. He found in a thrombosed capillary vessel, in a glomerulus in which no inflammatory changes had as yet occurred, numerous bacilli. They were also found in the epithelial cells of a convoluted capillary vessel. He was also able to demonstrate that the tubercular nodules in the kidney occurred around the capillaries in the glomeruli, often including the latter completely. He is of opinion that the bacilli migrate from the glomeruli into the surrounding tissues, giving rise to miliary nodules.

In cases of tuberculosis of the kidneys the number and peculiar arrangement of the bacilli in the urine are almost characteristic of this disease. Monpurgo[4] found, in the urine of a woman suffering from renal tuberculosis, numerous bacilli arranged in groups resembling the letter S (see Fig. 20)—an appearance only observed otherwise in pure cultures. Koch described this bacteriological condition in a case of miliary tuberculosis. For such a culture to form in the genito-urinary apparatus, it is necessary that the bacilli should be located in a place where they are not washed away by the urine and where they find a favorable soil for their growth—

---

[1] "Vorkommen der Tuberkelbacillen im Harn." *Centralbl. f. die med. Wissenschaft*, 1883, No. 5.

[2] *Fortschritte der Medicin*, B. i. p. 4.

[3] "Les bacilles dans la tuberculose miliaire. Tuberculose glomerulaire du rein." *Archiv de Physiol.*, 1886.

[4] *Schmidt's Jahrbücher*, B. ccxii. p. 128.

conditions that are furnished only in the kidneys. Philipowicz[1] found the bacillus of tuberculosis not only in three cases of tubercular pyelonephritis, but also in cases of acute miliary tuberculosis. If the bacilli were not present in sufficient number for detection by the microscope, their presence in the urine could be ascertained by the injection of the infected urine into the peritoneal cavity of guinea-pigs.

Meyer[2] found, in cases of miliary tuberculosis of the kidney affecting mainly the medullary portion, numerous bacilli in the uriniferous tubules, and, as he could exclude their entrance from that vicinity, he was of the belief that they were derived from the glomeruli, and he regarded the disease as an eliminative product. Tubercle bacilli do not pass through normal glomeruli, but no serious changes are requisite to permit their passage.

These observations go to show the frequency with which tubercle bacilli are found in the urine of patients suffering from renal and genital tuberculosis, and the manner of invasion through the general circulation. Experiments on the lower animals only confirm what has been related above.

Lentz[3] demonstrated the infectiveness of the urine of tubercular persons by inhalation-experiments on 5 rabbits. The animals were confined in a box into which the vapor of an atomizer introduced was mixed

---

[1] "Ueber das Auftreten pathogener Microörganismen im Harne." *Wiener med. Blätter*, 1885, No. 22.

[2] *Virchow's Archiv*, B. cxli.

[3] "Experimentelle Untersuchungen über die Infektiosität des Blutes und Urins Tuberculöser." Dissertation, Greifswald, 1881.

with the urine of two phthisical patients. For each two of the animals at one sitting 30–40 cubic centimeters of urine were vaporized. In 2 of the animals the experiment was continued daily for seventy-one consecutive days. The animals were then killed, and numerous tubercles were found in the lungs and liver. In 3 of the animals the urine was allowed to decompose under a cover of filtering paper, when it was administered by the same method. One of the animals died, in consequence of an abortion, on the ninth day; one was killed on the forty-eighth, the other on the sixty-ninth day. In all the animals well-marked tubercular processes were found in the lungs and the bronchial glands. The inhalation of urine from healthy persons proved harmless.

Borrel[1] has studied experimentally the etiology of renal tuberculosis, injecting pure cultures of the bacillus of tuberculosis into the vein of the rabbit's ear. In the majority of cases bacilli did not pass the pulmonary filter. To produce renal tuberculosis, cultures were injected into the aorta itself by way of the common carotid artery. Borrel concludes that there are two distinct forms of renal tuberculosis, differing less in their structure than in their origin. In cases of direct inoculation into the arterial system the tubercular process begins immediately after the inoculation. The bacilli that become arrested in the capillary vessels of the kidney are enveloped in epithelioid cells with abundant protoplasm and large nuclei, but without much chroma-

---

[1] *Annales de l' Institut Pasteur*, Feb., 1894.

tin; these cells going to form masses surrounded by smaller cells with less protoplasm and with nuclei rich in chromatin. These masses are strictly interstitial, the renal epithelium remaining free and undergoing changes in nutrition only secondarily. The tubercles are localized in the glandular and cortical substance. Borrel gives to this form the name of primary renal tuberculosis. In the second form, which commences about twenty days after interstitial inoculation and the formation of a distinct caseous center, and which Borrel calls granular tuberculosis of the kidney, invasion takes place through the lymphatics. In such instances there is more precise localization, the tubercles occupying the medullary and cortical substance. Here also, however, the process remains interstitial, the epithelium not being at all involved. These results differ altogether from the opinion of Baumgarten upon the participation of the fixed cells and renal epithelium, and confirm the views of Virchow and Metschnikoff, that tubercle is an accumulation of lymphatic cells, and that lymphatic granulation is interstitial in all organs.

Albarran[1] has recently made experiments for the purpose of producing renal tuberculosis in rabbits. He injected into the ureter a pure culture of the tubercle bacillus in the direction of the kidney, and then ligated the ureter at the point of puncture. Following such a procedure a number of tubercle nodules were found in the kidney, pelvis, and calices. Albarran states that it is easy to follow under the

---

[1] *La Semaine Médicale*, May 27, June 20, 1892.

microscope the ascending progress of the bacilli along the canals, and the alterations they cause. One other interesting result was the existence, along with the tubercular changes, of a diffuse nephritis, partly catarrhal and partly interstitial, with thickening of the vascular walls. Besides the tubercular changes localized in the kidney, the process extended to the mediastinal glands, the hip-joint of the same side, and the kidney of the opposite side.

These experiments go to show the danger to which the kidney is exposed, in persons suffering from pulmonary phthisis or tuberculosis of any other organ, by infection through the blood. Infection under such circumstances is very liable to give rise to bilateral miliary tuberculosis; but in many cases the process is limited to one organ, and in the absence of general miliary or advanced pulmonary tuberculosis the disease will pursue a chronic course, and is often amenable to successful surgical treatment.

Blood-infection results more frequently in tuberculosis of the parenchyma of the kidney than in tubercular pyelitis. Clinical experience and the result of post-mortem examinations leave no further doubt that infection through the blood not infrequently takes place, resulting in primary renal tuberculosis without an antecedent demonstrable tubercular focus in any other organ. These are the cases in which the most satisfactory results follow early radical operative intervention. In other cases the renal affection follows in the course of a latent focus which in itself does not impair the general health, such as cheesy bronchial glands, or latent joint- or bone-tuberculosis.

From these remarks it will be seen that the classification of renal tuberculosis from a practical standpoint made by du Pasquier[1] is no longer applicable to all cases. According to du Pasquier, tuberculosis may invade the kidney through the general circulation, or that organ may be reached by an ascending tubercular process through the ureter. In the first instance, he asserts that the process is limited to the cortical layer; in the second, the tubercles develop at the apex of the pyramids and become generalized through the organ, and to these modes of infection he makes correspond the two types of tubercular kidney, medical and surgical. In opposition to this view, it may be stated that the kidney and its pelvis are not infrequently the seat of a primary localized tubercular affection that comes within the range of legitimate and successful surgery; while, on the other hand, an ascending tubercular process may involve both organs at the same time, and, with its associated urinary or genital complications, may cause renal lesions of such gravity as to preclude absolutely the propriety of surgical interference. Although authorities differ greatly on the subject, it is safe to make the assertion that in nearly 50 per cent. of all cases of renal tuberculosis in men the kidney is reached secondarily by an ascending tuberculosis from the bladder and the genital organs. In women a primary vesical tuberculosis frequently gives rise to secondary tuberculosis of the kidney, which often constitutes in such cases the immediate cause of death. Secondary tuberculosis of the kidney fol-

[1] *Thèse de Paris*, 1894.

lowing a primary tuberculosis of the bladder is almost constantly preceded by an ascending tubercular ureteritis. Tubercular pyelitis following an ascending urinary tuberculosis is followed by an interstitial tubercular nephritis which usually has its starting-point at the papillæ, and reaches the parenchyma of the kidney by progressive extension along the lymphatic vessels and blood-vessels. Infection by this method usually involves both kidneys simultaneously or in succession after an interval of varying duration. Ascending tuberculosis of the urogenital tract is very frequently complicated by secondary infection with pus microbes—an event that invariably hastens the dissemination of the tubercular process and aggravates the local and general symptoms. Tubercular infection of the kidney from without as an isolated affection does not occur, or, to say the least, must be exceedingly rare.

Stintzing[1] states that primary renal tuberculosis is caused most frequently by cohabitation with a tubercular individual. Infection by coitus of the genital organs, both male and female, and of the lower portion of the urinary tract occasionally occurs; but infection of the kidney from such a source without a tubercular affection of the genitals or the lower part of the urinary tract is almost beyond the range of possibility.

The kidney may in rare cases become infected from a tubercular focus in its immediate vicinity. A specimen of such a rare method of infection is

[1] *Correspondenzblatt des allg. Ärzt. Vereins von Thüringen*, 1892, No. 8.

referred to by Hans Schmid[1] (see Fig. 24). The specimen is in the Pathological Museum at Greifswald, and illustrates the extension of a tubercular spondylitic abscess to the adipose capsule, finally reaching the kidney itself, resulting in perforation of the pelvis, tubercular pyelitis, nephritis, and ureteritis. A similar method of infection might occur in connection with retroperitoneal tubercular glands, tubercular ulcers of the intestines, and tubercular empyema.

Age must be regarded as a predisposing cause in the etiology of renal tuberculosis and in determining the character of the tubercular process, as appears from the statistics gathered by Guyon and Morris. Guyon[2] states that childhood is an age more or less predisposed to tuberculosis of the kidney; that, after children, puny adults are most frequently affected; and that of adult life, the ages between twenty and forty years are usually those manifesting the disease. He is also of opinion that renal tuberculosis occurs more frequently in men than in women, and is in the former not rarely associated with tuberculosis of the genital organs. According to this observer, tuberculosis occurs in man in two forms—the acute miliary and the chronic form—just as in other parts of the body. The former variety is rather more frequently met with in children than at any other age, and the kidneys are less often invaded than the other organs. The disease usually

---

[1] " Behandlung der Erkrankungen der Nieren und des Harnleiters (ausschliesslich der diffusen Nierenerkrankungen)." *Handbuch der Speciellen Therapie inneren Krankheiten*, B. vi. p. 305.

[2] *Wiener med. Presse*, 1890, Jan. 6–20; Sajous' *Ann. Univ. Med. Sci.*

involves both kidneys, and is found in the cortex as miliary granuloma. In the chronic form, which is more a disease of adult and middle life, the parenchyma is frequently the seat of marked degenerative changes, in which the calices and pelvis are often implicated.

Camargo[1] believes that tuberculosis of the urinary organs begins, as a rule, in the kidney, and by a descending process reaches the remaining portion of the urinary tract. In 25 post-mortems of chronic tuberculosis of the kidney made in the Geneva Hospital, he found in 25 tuberculosis of the bladder, and only in 11 tuberculosis of the genital organs; in 14 the genital tract was found free from tubercular affections of any kind.

Cayla,[2] in a general review of renal tuberculosis, asserts that this condition is secondary to a primary focus of tubercular infection in some other organ or tissue—lungs, skin, digestive or genito-urinary mucous membrane. He does not regard with favor the theory of infection through the last path by sexual contact, etc., opposing to this view his clinical observations and a series of experiments performed by him upon animals by injecting into the anterior part of the urinary tract pure cultures of the tubercle bacillus without causing, within fifty days, any tubercular manifestations in the parts. The histological structure of the anterior portion of the urethra, in his view, militates strongly against

---

[1] "Recherches anatomiques sur l'hyperplasie unilaterale d'un rein par suite d'aplasie de l'autre, et sur la Tuberculose descendante de l'appareil urinaire." *Revue Méd. de la Suisse Romande*, No. 10, 1892.

[2] *Gaz. des Hôp.*, Feb. 4, 1888.

this method of infection. He regards the renal infection as secondary to tuberculosis elsewhere, and usually transferred to the kidney by the hematic paths. The glomerules are the first to show invasion, later the tubules; finally the tuberculosis, by downward invasion, attacks the lower urinary apparatus.

James Israel[1] is firmly convinced, from his clinical observations and the result of post-mortem examinations, that primary tuberculosis of the kidney is a much more frequent affection than is generally supposed, and that tuberculosis of the bladder and the genital organs occurs in the majority of cases in consequence of a descending tubercular process. In one of his earlier cases a nodular swelling appeared in the region of the right kidney long after the patient had suffered from symptoms of a chronic cystitis. As the bladder-symptoms did not seem to warrant a diagnosis of vesical tuberculosis, he extirpated the right kidney and found a typical tubercular pyonephritis. The bladder-symptoms gradually disappeared after the operation. Israel presented specimens from three post-mortem examinations of patients who had died of tuberculosis of the urinary organs, and demonstrated that in every instance the disease had its primary starting-point in the kidney, followed by a descending tubercular process which eventually reached the bladder and the genital organs.

Donnadieu[2] quotes 29 reported cases of tuberculosis of the kidney to illustrate how little may be

[1] "Ueber Nierentuberculose." *Deutsche med. Woch.*, No. 31, 1890.
[2] *Arch. Clin. Bordeaux*, Nov., 1892.

gleaned from autopsy as to the direction of the infection in such instances—21 of the cases exhibiting the vesical and renal disease coexisting in such a state as to be separable; 1 case where both kidneys were normal and the bladder was much diseased; 2 cases where the bladder was apparently free from disease when the kidneys were affected; in 5 cases the records were so imperfect as to be unavailable. He also states that in one case, where there were no gross lesions of the bladder, microscopic examination indicated the existence of tuberculosis. Moreover, he quotes Guyon, who in 18 cases encountered 6 showing tubercular cystitis without renal involvement, and who has met but a single instance of unilateral renal tubercular lesion without other genito-urinary tuberculosis. Anatomical study being of such indifferent result, Donnadieu emphasizes the value of clinical observation in the recognition of the real progress of the process.

Examination of a number of clinical reports of cases regarded as examples of urinary tuberculosis indicates clearly that the kidney-symptoms rarely attract notice until after, and often long after, vesical symptoms are prominent. In many of the cases the prostatic body and the deep urethra are among the earliest to show evidence of involvement; and whether the kidneys are or are not the first portion of the urinary tract to become tubercular, it is certain that they are the last to show symptoms of their diseased condition.

Morris, in the records of the Middlesex Hospital, found the particulars in reference to age as an etiological element in 44 cases of renal tuberculosis to be as follows: Of 29 cases of miliary tuberculosis, 12 cases occurred in persons under ten years of age; 8 cases in persons between ten and thirty years; 6 cases in persons over thirty years; 3 cases in persons where age is not stated. Of 15 cases of the chronic form, 5 cases were between eleven and thirty years; 7 cases were over thirty years; 3 cases were in persons whose age is not stated. These figures indicate that miliary tuberculosis of the kidney is much more frequent in children than in adults, while the reverse is the case in persons over twenty years of age, who are more frequently affected with the chronic form. Surgical treatment of renal tuberculosis is therefore much more frequently indicated and required in adults than in children.

In the chronic form of renal tuberculosis Steinthal[1] found that in 12 out of 24 cases the disease was limited to one kidney.

Tuberculosis of the urinary organs is much more frequent in men than in women, owing to the fact that in men genital tuberculosis frequently extends to the urinary tract, while in women the genital and urinary organs are two separate systems, and genito-urinary tuberculosis in them can be the result only of a simultaneous infection of both systems or of the formation of fistulæ between them—conditions that seldom present themselves.

Symptoms and Diagnosis.—The symptoms of tuberculosis of the kidney are insidious in their onset and ill defined in their clinical aspects. None of

---

[1] *Virchow's Archiv*, B. c. p. 81.

the symptoms are pathognomonic, and a probable or positive diagnosis can be made only by a careful study of the clinical history and by recourse to all diagnostic means, including a careful bacteriological examination of the urine. The symptoms will vary according to the form of tuberculosis and the presence or absence of complications. In miliary tuberculosis the symptoms referable to the kidney-lesion are often slight or entirely absent, and the diagnosis is more often made in the post-mortem room than at the bedside. If renal or vesical symptoms make their appearance in cases of pulmonary or genital tuberculosis, tuberculosis of the urinary organs should be surmised and the necessary investigation made. Symptoms of chronic cystitis, in the absence of a local cause for the supposed inflammation of the bladder, should induce the practitioner to search for evidences of a renal affection, as the first symptoms of renal tuberculosis are often referred to the bladder. A remarkable instance of this kind has been recorded by Huber.[1] The patient had suffered for some time from pain in the neck of the bladder and in the urethra and from frequent micturition. The urine was at first pale and contained no albumin. There was no change detectable in the bladder, but the use of a catheter caused great pain. Some time afterward pus appeared in the urine, micturition became extremely frequent—perhaps every quarter of an hour—and the urine occasionally contained a little blood. There was no pain over the symphysis pubis. The patient died

---

[1] *Deutsches Arch. f. klin. Med.*, B. iv. p. 609.

in a condition of exhaustion, and on post-mortem examination cavities and caseous masses were found in both lungs. The kidneys, especially the left one, were much enlarged. The structure of the right kidney appeared normal, but the left one contained numerous ulcerating cavities opening into the pelvis of the kidney, and much caseous material. The bladder was contracted; its coats were hypertrophied; its mucous membrane and that of the urethra were infected, but presented no trace of caseous material. The sexual organs were intact. In this case all symptoms were referred to the bladder, which was comparatively normal.

In one of Trendelenburg's[1] cases the renal tuberculosis was so overshadowed by vesical symptoms that the bladder was opened above the pubes, and, as a small ulcer was found near the urethral opening, a considerable portion of the bladder-wall was resected. The operation was followed by relief. No tubercles were found in the specimen removed. Soon there appeared symptoms that pointed to the left kidney as the seat of the disease. Nephrectomy was performed. The pain in the bladder returned soon after the second operation with its former severity. In the kidney numerous tubercular foci of considerable size were found. This case goes to show that renal tuberculosis may remain in a latent condition for a long time, manifested by a non-tubercular cystitis.

In miliary tuberculosis the local symptoms are so masked by those of the primary disease or by

---

[1] "Perthes, ueber Nierenextirpation." *Deutsche Zeit. f. Chir.*, B. xlii. Heft 3.

the diffuse general miliary tuberculosis that during life the existence of the renal complication is not even suspected. In a few of the cases of this kind that have come under the observation of the writer the most prominent symptoms that could be referred to the kidneys were lumbar pain, frequent desire to urinate, occasionally slight hematuria, and traces of albumin. Tubercle bacilli were found in the urine in only about one-half of the cases. In the case of a boy fourteen years of age, in whom the disease was limited to the urinary system, the infection in the course of a year extended as far as the meatus. The urethritis was quite acute and was attended by profuse suppuration. Urine drawn from the bladder contained numerous tubercle bacilli. In the urethral discharge pus microbes were found. The acute urethritis subsided in the course of four or five weeks, and with the subsidence of the acute inflammation the purulent discharge ceased. The most prominent clinical feature during this stage of the disease was almost complete incontinence of urine.

In chronic tubercular nephritis and pyelitis the symptoms differ somewhat in the ascending and descending forms. In the ascending form the renal affection is preceded by symptoms referable to a chronic inflammation of the lower portion of the urinary tract, and when the pelvis of the kidney is reached the urine from the very beginning contains the characteristic products of tubercular inflammation. It is also in this form that the kidney is found enlarged—the result of retention of urine by ureteric obstruction; and the retention gives rise to periodical attacks of renal colic. In primary renal tuberculosis the extension of the inflammatory process to the ureter, bladder, and lower urinary tract is strong, if not positive, proof of its tubercular nature. Such extension seldom occurs in calculous and suppurative pyelitis.

If in men the genital organs are involved during the course of the disease, such complication is another proof of the tubercular character of the primary renal affection.

We are indebted to Ammon (1833) for the first attempts to formulate and describe the symptoms caused by tubercular nephritis and pyelonephritis. Caseous inflammation of the pelvis of the kidney should always be suspected when renal symptoms develop in men the subject of tuberculosis of the genital organs. An examination of the prostate is absolutely necessary in all cases of pyuria, and might here also give a clue to the diagnosis. The early loss of strength and signs of a rapidly progressing marasmus should excite suspicion of the tubercular nature of the kidney-affection. A careful examination of the complexus of symptoms presented by chronic tubercular nephritis and pyelonephritis is absolutely necessary, and it is only upon such a basis that a probable or positive diagnosis can be made in the absence of well-marked tuberculosis in any other organ.

Pain.—In acute miliary and chronic tubercular nephritis pain is slight or is entirely absent. Pain often becomes a distressing symptom when the tubercular foci rupture into the pelvis and when the pyelonephritis is caused by an ascending tubercular ureteritis. In the latter instance the pain results from obstruction to the free escape of urine and tubercular products. In such cases the pain is severe, paroxysmal, and extends along the ureter and the lower part of the urinary tract as far as the meatus and the testicle, which retracts during the attacks. The pains simulate so closely those produced by stone in the bladder that such cases are frequently subjected to repeated instrumental examination of the bladder in fruitless attempts to detect a stone. In some cases the pain is dull, of slight intensity, and referred to the lumbar region; in others the pain is exceedingly sharp and severe. Occasionally the pain also extends along the inner surface of the thigh. The character of the pain should be carefully taken into consideration in the differential diagnosis between calculous pyelitis, stone in the bladder, and tubercular nephritis.

Tenderness.—As an indication of renal disease, tenderness is of diagnostic value in the differential diagnosis between inflammatory affections of the kidney and tumors, in cases in which the kidney is sufficiently enlarged to render it accessible to palpation. This symptom is of little or no value in the diagnosis of acute miliary tuberculosis of the kidney, as the organs, as a rule, cannot be felt by bimanual palpation. If the affected kidney can be palpated, tenderness can always be elicited in the tubercular affections, but it is not so acute as in cases of suppurative pyelitis or pyonephrosis. The pain produced by pressure often extends along the ureter to the bladder and urethra.

Strangury.—Strangury frequently exists without vesical tuberculosis, as in Huber's case, and is then caused by the altered condition of the urine. Morris mentions the case of a man under his care who passed urine with much pain and spasm on an average 160 times in the twenty-four hours. At the post-mortem examination there was found chronic tuberculosis of the kidneys and ureter, the bladder having only quite recently become affected. Tubercular pyuria seldom exists for a considerable length of time without some evidence of bladder-irritation.

Faure[1] mentions a case in a man aged twenty-four years, who, upon entrance into the hospital, showed symptoms of an intense cystitis characterized by distressing strangury, and at the autopsy, a year or so later, the prostate was found involved, the bladder-walls thickened and ulcerated, the right ureter thickened, ulcerated, and the seat of cheesy deposits at several points, and the kidney on the same side converted into a tubercular abscess. The disease in this case had its origin in the kidney, and the distressing vesical symptoms appeared long before the infection reached the bladder.

These cases show the necessity of careful methods of examination in cases of renal tuberculosis manifested by early vesical symptoms, as repeated examination for stone in the bladder has often been the means of inflicting an incalculable amount of harm in such cases, and many mixed infections can be traced to such harmful procedures.

[1] *Journ. Am. Med. Assoc.*, 1890.

Fever.—Increased temperature always attends miliary tuberculosis of the kidney, and more or less fever is generally present in the chronic form of the disease. The temperature depends often more on the primary disease than on the secondary renal affection. Prescott and Goldthwait,[1] in reporting 4 cases of renal tuberculosis, all with a fatal result, call attention to two important symptoms—variability in the appearance of the urine and a subnormal temperature. In 3 of the cases the temperature was subnormal, notwithstanding that suppuration was profuse. In two young adults, a male and a female, that were under the observation of the writer for several months, while suffering from tuberculosis of the entire urinary system, the temperature was either normal or slightly subnormal in the morning, while the evening temperature was invariably from 100° to 102° F.

Swelling.—The kidney is seldom palpably enlarged in miliary tuberculosis. According to Newman,[2] in about one-fifth of the cases recorded, and in the majority operated upon for the chronic form, the tubercular pyonephrosis was discoverable by palpation during life. Increase in the size of the kidney, while it must be regarded as an important evidence of disease, is one which, when present in connection with a tubercular lesion, usually indicates that the disease has so far advanced as to cause ureteric obstruction. The formation of a perinephric abscess increases the size of the swell-

---

[1] "Observations on Tuberculosis of the Kidney, with a Report of Cases." *Boston Med. and Surg. Journ.*, Jan. 15, 1891.
[2] *On the Diseases of the Kidney amenable to Surgical Treatment*, p. 291.

ing. Perforation into the intestines or into the peritoneal cavity is exceedingly rare.

The most important and reliable diagnostic information in renal tuberculosis is furnished by a careful examination of the

Urine.—In all cases of suspected renal tuberculosis instrumental examination should be preceded by a careful analysis and microscopic examination of the urine. Observance of this rule will frequently be the means of preventing painful and so often harmful instrumentation. In miliary tuberculosis, and in the chronic form during the beginning of the disease, the quantity of urine is usually increased. This temporary polyuria is the result of stimulation of the glandular tissue, caused by the increased vascularity induced by the presence of miliary nodules that have not advanced sufficiently far to compromise the parenchyma of the organ. Before destructive processes commence within the renal pelvis or substance, traces of albumin, or even a few red blood-corpuscles, may be detected; but when the pelvis is primarily affected, or the renal foci have ruptured into the pelvis, the urine invariably contains *débris* of tubercular tissue, and the albumin and blood are increased in quantity. Osler[1] states as an aid in the differential diagnosis in pyelonephritis and tubercular nephritis, that the urine remains acid in renal tuberculosis unless there is an extensive coexistent cystitis. Tubercular urine is albuminous and is mixed with blood, pus, and *débris* of tubercular tissue. It rarely contains casts. Hema-

[1] *New York Med. Journ.*, July 26, 1894.

turia is seldom profuse; but pyuria is constant when the ureter has not been obstructed, and intermittent in partial obstruction—important features in the differential diagnosis between renal tuberculosis and calculous pyelitis. In the latter hematuria is often profuse and pyuria is intermittent. In rare cases of renal tuberculosis hematuria is profuse and becomes a menace to life.

Trautenroth[1] observed a case of renal tuberculosis in which profuse hemorrhage made it necessary to extirpate the kidney sixteen days after the appearance of the first symptoms. The patient suffered from severe pain in the right lumbar region. She was a single woman, twenty-four years old, the subject of incipient pulmonary tuberculosis. The pelvis of the kidney contained coagula, and upon one of the papillæ, greatly enlarged, was an ulcerated surface. A more careful examination of the organ showed diffuse tuberculosis. In the urine removed before the operation one tubercle bacillus was found. The wound healed promptly, and the patient made a good recovery.

Routier[2] observed a case of tuberculosis of the kidney in a patient twenty-eight years old, in which the first symptom was quite severe hemorrhage which continued for seventeen days, followed by renal colic and frequent desire to empty the bladder. Disease of the right kidney was diagnosed by the use of the cystoscope. On making pressure over

---

[1] "Lebensgefährliche Hæmaturie als erstes Zeichen beginnender Nierentuberculose." *Centralb. f. Chir.*, 1896, No. 16.

[2] "Tuberculose renale, Hæmaturie Guérison." *Bull. et mém. de la Soc. de Chir. de Paris*, vol. xxi. p. 148.

the right kidney blood escaped from the ureter. A probable diagnosis of tumor was made, and six weeks from the commencement of the disease the kidney was removed through a lumbar incision. Although the patient was extremely anemic at the time of operation, he recovered promptly. On examination of the kidney it was found that the tubercular process was limited to two calices, which were the seat of ulceration.

Examination of the sediment of tubercular pus under the microscope shows epithelial cells from the different infected territories, tubercular detritus, blood, and, above all, tubercle bacilli. Stintzing regards the absence of tube-casts in chronic renal affections as characteristic of renal tuberculosis. For the purpose of hastening sedimentation of the urine it is advisable to make use of Stenbeck's sedimentator; this is especially necessary in cases in which the urine contains but few morphological elements which furnish on standing no distinct deposit. If such an instrument is not at hand, the urine should be allowed to settle for several hours in a champagne-glass, after which the clear fluid is poured off and the sediment examined. When the renal affection is complicated by inflammation of a bladder the seat of a mixed infection, the urine is so much more turbid and viscid, in consequence of the masses of bloody mucus it contains, that it is voided with difficulty. In such cases it is advisable to treat the urine as advised by Lentz for the purpose of eliminating from it the substances that interfere with a satisfactory microscopic examination. "The demonstration of small and occasionally of larger cheesy fragments, which are insoluble in acetic acid, and consist microscopically of inspissated cheesy elements, of granular detritus, and elastic fibers, is of great pathognomonic importance" (Newman). The morphology of the epithelial cells contained in the urine will indicate the location of the tubercular process. The presence of renal epithelia would point to the kidney as the seat of the disease, while epithelial cells from the pelvis, ureter, and bladder would be an indication as to the extent of the descending infection. The bacteriological examination of the urinary sediment is of the most important diagnostic value, and should always be supplementary to the microscopical examination of morphological elements, as it alone furnishes the most reliable evidence upon which to base a positive diagnosis. In suppurative pyelitis the urine contains one or more varieties of pus microbes; in tuberculosis of the kidney the tubercle bacillus can often, but not always, be demonstrated in the urine, while pus microbes are also present in tuberculosis of the urinary tract complicated by secondary mixed infection with pus microbes.

Staining of Bacilli.—Excellent directions for staining bacilli in tubercular urine are given by J. W. White in Dennis's *American System of Surgery:* "To examine a specimen of urinary sediment for tubercle bacilli it must be thinly smeared upon a cover-glass and dried, and then floated, film-side down, on a solution of anilin-magenta or gentian-violet; when quite deeply stained the cover-glass is rinsed in a 25 per cent. solution of nitric acid to decolorize everything but the bacilli. The acid is then washed off in distilled water, and the slide again floated, film-side downward, in a solution of methylene-blue, in order to counterstain the tissues. The bacilli will now appear red in a blue field."

If tubercle bacilli are not present in the urine, the coexistence of tubercular affections in other organs renders the diagnosis of renal tuberculosis probable if not positive, as well as the existence of fever of a hectic type. Tuberculosis must be assumed if renal epithelia and fragments of tubercle-tissue are found in the urine.

In the diagnosis of renal tuberculosis systematic examination of all the organs predisposed to tuberculosis, and careful inquiry into the history of the case, should be attended to before the symptoms referable to the kidney-lesion are considered separately and collectively. The general health of patients suffering from renal tuberculosis is usually moderately impaired at an early stage of the disease, while it often remains unaffected in pathological conditions giving rise to enlargement of the organ, as in hydronephrosis, echinococcus, and pyonephrosis. The marasmus and hectic fever increase with the advance of the disease. Appetite and digestion are impaired. Sleep is disturbed by pain and the frequent desire to evacuate the bladder. The vesical distress is often great without a coexisting vesical tuberculosis. Retinitis albuminurica, so often present in the different forms of inflammatory and degenerative affections of the kidney, is absent in renal tuberculosis. In advanced bilateral renal tuberculosis, and in unilateral disease in case the opposite organ does not undergo an adequate hypertrophy, uremic symptoms, such as headache and vomiting, mark an advanced stage of the disease. In the absence of palpable swelling and of tubercular products in the urine the diagnosis is exceedingly unsatisfactory, even after a most thorough examination. In such cases all diagnostic resources must be employed to enable the practitioner to make a probable diagnosis by exclusion and, if need be, by the results obtained from inoculation-experiments.

Palpation.—The healthy kidney in its normal location can seldom be felt by bimanual palpation—a statement which conflicts with the experience of James Israel. Practically, a palpable kidney is either a movable or an enlarged kidney. Renal palpation (Fig. 25) is a difficult and often unsatisfactory task in obese and nervous patients, owing to the great thickness of the abdominal wall in the former instance, and rigidity of the abdominal muscles in the latter. The administration of an anesthetic may become necessary in those cases in which the muscular rigidity cannot be overcome by ordinary means. The bowels should be thoroughly moved by a cathartic or a high enema before palpation is practised, as fecal accumulation in the hepatic or splenic flexure of the colon might easily give rise to mistakes in diagnosis, or at least interfere with a satisfactory examination by this method. If the affected organ is enlarged sufficiently to be felt in the lumbar region and below the costal arch, bimanual palpation will determine its exact location, its size and form, and the existence of tenderness. The patient should be placed in the dorsal recumbent position, with the head and chest slightly raised and the leg and thigh flexed. The surgeon should sit upon the bed or couch, at the side of the patient, and place the tips of the fingers of the right hand, in palpating the right kidney, beneath the costal arch, and the left hand against the lumbar region. The patient should be asked to breathe in a natural way, and during respiration the right hand is gently pushed in the direction of the lumbar region while firm pressure is

FIG. 25.—Renal Palpation.

made against the lumbar region with the left hand. If the kidney is enlarged sufficiently, it can be felt during inspiration between the two hands, and its consistence, form, and size ascertained. If the kidney projects freely beneath the costal arch, it can readily be palpated, and in pyonephrosis fluctuation can often be distinctly felt. If palpation as first described does not prove successful, the patient should be placed on the opposite side and the procedure repeated. In some cases palpation proves most successful with the patient in the sitting or standing position. If the renal tuberculosis is complicated by a perinephric tubercular abscess, the swelling is well marked, fixed, and fluctuates on palpation.

Percussion.—The renal swelling can often be outlined satisfactorily by percussion. The patient should be placed in the ventral recumbent or the erect position. The lumbar dulness is a fair indication of the location and size of the kidney. Anterior percussion will locate the colon in front of the renal swelling.

Rectal Insufflation.—It is not always easy to differentiate between an enlarged kidney and an intra-abdominal swelling, such as a distended gall-bladder, an echinococcus cyst of the liver, or an intestinal tumor. To distinguish an intra-abdominal from a retroperitoneal swelling rectal insufflation is of the greatest diagnostic importance. The insufflation of air should be made with a large rubber bag holding at least a volume of air equal to four gallons, supplied with a stop-cock near the bag, to which is attached a piece of rubber tubing 4 to 6 feet in length, with the tip of an ordinary vaginal syringe securely fastened into the opposite end. The patient is placed in the dorsal recumbent position. One assistant with one or both hands presses the margins of the anus against the rectal tube, to prevent the escape of air, while a second assistant makes the necessary pressure by sitting upon the rubber bag placed on a chair, when the stop-cock is turned on and the inflation made slowly and without interruption. By placing the hands over the sigmoid flexure the entrance of air can be distinctly felt. The insufflation should be made very slowly, and should be continued until the cecum is well distended. Having ascertained and outlined previously the dull area on percussion, the effect of insufflation is now ascertained by repeating the percussion. If the swelling is retroperitoneal, the dulness disappears, owing to the distended tympanitic colon in front of it; if it is intra-abdominal, it may become displaced or diminished by the overlapping colon, but it does not disappear entirely.

Rectal Palpation.—Simon[1] recommends manual exploration of the rectum as a reliable diagnostic resource in differentiating between renal and pelvic swellings and tumors, and in ascertaining the conditions of the ureters. This method of examination requires the use of an anesthetic, and should never be undertaken by surgeons with large hands. It is inapplicable in the case of children and persons with a small pelvis.

Catheterization of the Ureters.—This method of examination is especially useful in cases of suspected renal tuberculosis in women. It enables the surgeon to secure urine from the pelvis of the kidney for examination, and in this way obtain positive evidence of the condition of both organs—a matter of greatest consequence when the question of operative interference arises. To Simon belongs the credit of having devised this exceedingly valuable

[1] *Chirurgie der Nieren.* Erlangen, 1871, p. 206.

diagnostic aid. In his first attempts he succeeded in 15 out of 17 cases—a success which it would be difficult to duplicate to-day, notwithstanding the great improvements in the technique made by Pawlik and Kelly. Simon dilated the urethra sufficiently to insert the left index finger, which served as a guide for the catheter. As the most important intravesical landmark he took first the ligamentum interuretericum. Catheterization of the ureter from the pelvis of the kidney he considered possible only in cases in which the pelvic orifice of the ureter is dilated. Later, in catheterization of the ureter from the bladder, he used the cervix uteri as a guide. The orifice of the ureter is ¾ to 1 centimeter to the outside of the commissure of the cervix, and ¼ centimeter in front. Those who have ever seen Kelly catheterize the ureters must have been astonished at his dexterity and the perfection of his instruments and technique. Anyone who inclines to become an expert in catheterization of the ureters must study his directions and secure his set of instruments. This part of the diagnostic work should be entrusted to experts, as few surgeons will acquire the necessary dexterity in the use of the instruments.

Catheterization of the ureters in the male, through the intact bladder, by the use of the cystoscope, has been successfully performed by a few experts, but it is a procedure vastly more difficult than in women. Catheterization of the ureters through a suprapubic opening in the bladder has been done in a few instances, but it is by no means an easy task, and it will be resorted to only in exceptional cases.

Cystoscopy.—The cystoscope is useful both in men and in women in receiving the urine as it escapes from the ureters for chemical and microscopical examination, and in ascertaining whether or not the tubercular process is an ascending or a descending one. In descending infection the mucous membrane around the opening of the ureter on the affected side is often studded with miliary tubercles.

In a considerable number of cases all the diagnostic resources that have been enumerated do not suffice in demonstrating the tubercular nature of the renal affection. It is in such cases that, as a last resort, a positive diagnosis can be made by the results obtained from

Inoculation-experiments.—The bacilli in the tubercular urine may be so few or so much changed by the altered urine that their detection by staining and the use of the microscope is difficult or impossible. It has been shown experimentally that the injection of tubercular urine into the subcutaneous tissues or the peritoneal cavity of guinea-pigs or rabbits causes typical tuberculosis, while normal urine proves harmless and is removed in a short time by absorption. A few drops of the sediment or a hypodermic syringeful of urine should be injected into the subcutaneous tissue in the groin or into the peritoneal cavity, under strict antiseptic precautions. If the urine is free from tubercle bacilli, the result is negative; if the urine is from a tubercular kidney, the animal will die of acute miliary tuberculosis in the course of five or six weeks. If the injection is made into the subcutaneous tissue, a hard nodule will form in the course of a few days at the point of injection, and in two or three weeks tubercle lymphadenitis will develop. Later the disease may extend to the peritoneal cavity, before or about the time general miliary tuberculosis sets in. If the peritoneal cavity is infected, diffuse tubercular peritonitis is followed by general miliary tuberculosis, usually five or six weeks after the inoculation.

Prognosis.—The prognosis in renal tuberculosis is always grave. Bilateral miliary tuberculosis is followed by death in a short time from the primary affection, death being hastened by the diffuse general miliary tuberculosis. A spontaneous cure is possible, in cases of limited renal tuberculosis with involvement of the pelvis of the kidney, only when the disease becomes encapsulated and the cheesy material becomes inspissated or calcareous. Such a favorable termination must be exceedingly rare. Fibroid tuberculosis of the kidney has not been described, and it is probable that the local conditions surrounding the tubercular focus do not permit of such a favorable course of the disease.

Great thickening of the capsule of the kidney must be regarded as a favorable change, as it prevents extension of the tubercular process to adjacent tissues. The extension of the infection to the adipose capsule and the formation of a paranephric tubercular abscess hasten the fatal termination. A renal tuberculosis may remain stationary for several years under favorable local and general conditions. In all cases of renal tuberculosis in which the disease is progressive, as is usually the case, acute exacerbations occur during intervals variable in duration. Profuse hemorrhage in exceptional cases becomes a source of danger to life from loss of blood and from blocking of the ureter by coagula. Anuria may set in, even if the opposite kidney is healthy, in cases in which the ureter on the opposite side becomes suddenly obstructed by a coagulum, by fragments of tubercular tissue, or by calcareous masses. The most favorable cases, so far as the duration of life is concerned, are those in which the disease had its origin in the pelvis of the kidney and is limited to one side. These are also cases most amenable to surgical treatment.

As a rule, renal tuberculosis is characterized by its progressive nature. If the disease had its primary origin in the kidney, the pelvis is sooner or later involved, when the tubercular process, by continuity of surface, extends in a downward direction. Infection may extend from the kidney to the meatus, implicating the mucous membrane of the entire urinary tract. In ascending tuberculosis and primary tubercular pyelitis the parenchyma of the kidney is invaded in a short time, and, if life is sufficiently prolonged and the disease is limited to one side, results sooner or later in complete destruction of the organ.

Sufficient time has not elapsed to judge of the curative results of operative treatment. A number of cases of unilateral primary tuberculosis of the kidney treated by nephrectomy have been placed on record in which it is stated that the patients remained in good health several years after the operation. James Israel has had a larger experience in renal surgery and more satisfactory results than any other surgeon. Without timely surgical treatment in primary renal tuberculosis life is seldom prolonged beyond two or three years, while death occurs much sooner if the renal affection is dependent upon tuberculosis of another important organ.

Treatment.—Owing to the uncertainty of diagnosis during the early stages of renal tuberculosis, and the existence in more than 50 per cent. of all cases of incurable complications, the surgical treatment has not yielded as many permanent results as was anticipated during the early history of nephrectomy for this affection. It must be remembered, too, that extirpation of the kidney is a serious operation which is not to be lightly undertaken, and that patients suffering from renal tuberculosis seldom seek the services of a surgeon until serious complications have set in. The surgical treatment of renal tuberculosis will yield more promising results when physicians recognize the importance of making an early and positive diagnosis and encourage their patients to seek timely surgical aid. Again, the statistics of surgical intervention in renal tuberculosis will make a much better showing when surgeons take more pains in the selection of cases. It is just as important, in the interest of patients and the advancement of surgery, to seek for and study the contraindications as it is to study the indications for an operation. The operative treatment should be limited to cases in which the strength of the patient is sufficient to resist the immediate effects of the operation, and in which the renal disease either is primary or, if secondary, the primary disease is not far advanced. Palliative operations are indicated when a tubercular abscess forms. When the affection, especially if primary or localized, can be recognized, the remedy lies in resort to surgical measures, and, in accordance with the conditions present, these measures will consist principally in nephrectomy or nephrotomy.

The general treatment of renal tuberculosis should never be overlooked. The internal use of salol or benzoate of ammonium will often afford great relief when the disease is attended by much vesical distress. Bitter tonics and mineral acids are indicated when the appetite and digestion are impaired and the urine is alkaline and loaded with pus. The prolonged use of guaiacol or its carbonate, if well borne by the stomach, has a salutary influence in retarding the disease. Strangury should be relieved by warm sitz-baths and by opium and belladonna administered per rectum. Chills and high temperature call for quinine in 5- to 10-grain doses, repeated once or twice at intervals of half an hour. Alcoholic stimulants are contraindicated as a rule. Iced champagne is of value in the treatment of nausea and vomiting the forerunners of uremic intoxication. In threatened uremia warm baths or the pack and cathartics, especially calomel and salines, meet the symptomatic indications. A non-stimulating but nutritive diet, especially milk, constitutes an important item in the treatment of renal tuberculosis, as in other inflammatory and degenerative diseases of the kidney. A change of climate will often accomplish a great deal in retarding the disease in inoperable cases and in hastening recovery and preventing relapse after operation.

Direct treatment of tubercular pyelitis by catheterization of the pelvis of the kidney and washing it out with antibacillary solutions has not yielded very encouraging results. This method of treatment has been the means of causing a secondary infection of the tubercular territory, and after a fair trial will probably be abandoned. There recently came under the observation of the writer a case of unilateral tuberculosis of the pelvis of the kidney that had been treated by this method for two consecutive months. The pelvis of the kidney was washed out daily through a ureteral catheter without any appreciable benefit; on the other hand, the general condition of the patient was greatly impaired by the pain caused by the treatment. The diseased organ was then removed by lumbar nephrectomy. The wound healed promptly, and the general health of the patient improved steadily after the operation. There were no signs of recurrence or of tuberculosis in any other organ six months after the operation. The urine, which had been constantly turbid before the operation, and in which bacilli and fragments of tubercular tissue were found, cleared up immediately after the removal of the kidney, and remained clear and normal in quantity, showing that the opposite kidney was intact. The pelvis of the kidney was found covered with a layer of cheesy material, and in the substance of the kidney, near the pelvis, there were found a number of miliary tubercles which did not present any evidences of cheesy degeneration. The disease evidently commenced in the pelvis and was extending to the parenchyma of the organ. The vesical symptoms, which were a prominent feature in the case, disappeared soon after the operation.

The following case, reported by Casper,[1] illustrates the success attending early recognition and prompt radical treatment of tuberculosis of the kidney. The patient was a woman forty-two years old, without hereditary predisposition, who for six months had complained of symptoms of vesical catarrh, to which were added pains in the right loin, together with a sense of abdominal tension. Varied local treatment proved unsuccessful. The urine was slightly turbid and of acid reaction, with a specific gravity of 1018. It contained numerous pus-corpuscles and ciliated epithelium, but neither blood-corpuscles nor tube-casts. A small quantity of albumin also was present. In the centrifugated sediment tubercle bacilli were found. Upon cystoscopic examination the interior of the bladder appeared normal except in the situation of the right ureter, the opening of which could not be seen. In its usual site was a diffusely reddened elevated area in which blood-vessels could not be recognized. The urine from the left ureter appeared clear and normal, while none could be seen to issue from the place where the right ureter should have been, although from time to time a few drops collected in this situation. The left ureter was readily catheterized with the aid of the uretero-cystoscope, and the urine obtained proved to be normal. After a number of unsuccessful attempts to enter the right ureter the catheter disappeared beneath the mucous membrane of the bladder and was readily pushed further onward. The urine thus obtained resem-

[1] *Berliner klin. Woch.*, 1896, No. 17, p. 369; *Medical News*, May 30, 1896.

bled that found in the bladder, being turbid, purulent, and albuminous. Upon the first examination tubercle bacilli could not be found, but at a subsequent examination their presence was conclusively demonstrated. A diagnosis was therefore made of tuberculosis of the right kidney with circumscribed involvement of the bladder. Accordingly, nephrectomy was performed, with recovery of the patient. The removed kidney presented tubercles upon its surface and two caseous nodules in its parenchyma.

Operations for Tuberculosis of the Kidney.—The operative treatment of tuberculosis of the kidney is either palliative or radical.

Palliative operations are indicated in many cases as a preliminary treatment to radical measures, in others for the relief of pain and the removal of liquefied products of a large cheesy focus with or without secondary infection. The opening of a paranephric abscess that has developed in the course of a renal tuberculosis, and nephrotomy for a tubercular pyonephrosis, must be regarded in the light of palliative rather than curative measures; yet under proper conditions both these procedures prepare the way in some cases for a more accurate diagnosis and more successful radical treatment, and in other cases for successful local treatment. Palliative operations often prove of great value in checking the local extension of the disease and in retarding reinfection of the organism. Incision, scraping, disinfection, and draining of a paranephric tubercular abscess are well calculated to protect the adjacent tissues against invasion by the tubercular process. Incision and drainage of a distended tubercular pelvis of a kidney relieve tension and often accomplish more in subduing pain than the administration of narcotics, and certainly retard local and general infection. Nephrectomy must be limited to cases in which the surgeon can satisfy himself that the opposite organ is present and in a healthy condition. The selection of cases of unilateral renal tuberculosis for primary nephrectomy often requires much time and repeated examinations. Haste in coming to positive conclusions in regard to the feasibility and safety of an operation is often followed by unexpected disastrous results. The final decision must often be left in the hands of expert diagnosticians and professional bacteriologists. It is in such cases that consultations are in place, and should more frequently be requested by the attending physician.

The relative value and danger to life of nephrotomy and nephrectomy in the treatment of renal tuberculosis are well shown by the experience of Guyon,[1] as related in a lecture before the Faculty of Medicine in Paris. He announced the following statistics in 33 cases:

Nephrotomies (9 cases).

Died	2 (22 per cent.).
Cured	1 ⎫ two fistulæ.
Relieved	1 ⎭
Unknown	1
Under observation	4

Nephrectomies (24 cases).

Twenty primary	Died	11 (55 per cent.).
	Cured	7
	Relieved	1
	Unknown	1
Four secondary	Died	1 (25 per cent.).
	Cured	3

[1] *Ann. des Mal. des Organes Génito-urinaires*, Sept., 1889.

In a recent communication emanating from the Pepper Laboratory of Clinical Medicine, Hamill[1] considers renal tuberculosis in its various clinical aspects, and reports 55 cases that he has collected from the literature. Operation was performed in 17 cases of this number, with the following results:

	Cases.	Improvement.	Recovery.	Death.
Nephrotomy	4	1	2	1
Nephrectomy	9	1	5	3
Nephrotomy followed by nephrectomy	4	1	2	1

These results are certainly encouraging, and should induce physicians and surgeons to make early and faithful efforts to give this class of patients suffering from an otherwise incurable disease the benefits to be derived from early operative interference.

If an operation is decided upon, certain precautions in performing it should never be neglected. All patients suffering from renal disease necessitating an operation require special attention in the use of anesthetics. The choice of the anesthetic is a matter of great importance. With few exceptions, chloroform is preferable to ether, as it is less likely to produce serious vascular changes in the kidney. The anesthetic should be absolutely pure, and should be administered by the drop method. The patients should be prepared carefully for several days. The use of cathartics, a warm bath in the evening, and an exclusive milk-diet are the most important preparatory measures. Shock should be prevented by the administration of strychnine before the anesthetic

---

[1] *Primary Tuberculosis of the Kidney, with Especial Reference to its Manifestation in Infants and Children*, p. 35. Philada., 1896.

is given, and by surrounding the limbs and those portions of the body not exposed with warm blankets during the entire operation. Hemorrhage should be carefully arrested during the progress of the operation by the use of forcipressure-forceps. Patients suffering from renal tuberculosis are always badly affected by the loss of even a small quantity of blood. The defective urinary secretion calls for special caution in the use of certain antiseptics, such as carbolic acid, corrosive sublimate, and iodoform, owing to the increased risk of dangerous if not fatal intoxication. The writer had a painful experience in the use of iodoform in a case of renal tuberculosis complicated by genital tuberculosis and a tubercular paranephric abscess. The patient was a man about thirty-five years of age, who suffered from tuberculosis of the right testicle which terminated in the formation of abscesses and fistulæ. Later the disease extended, by way of the bladder and ureter, to the kidney on the same side, and shortly before the operation was performed a large paranephric abscess formed and presented itself in the lumbar region in the form of a well-marked prominence which was covered by edematous skin and fluctuated distinctly on palpation. The testicle was removed, more for the purpose of palliation than with a view to influence for the better the disease of the urinary organs. The operation was performed under partial anesthesia, as the patient was very anemic and considerably emaciated. The paranephric abscess was tapped with a small trocar, its contents evacuated, and about 3 drams of a 10 per cent. emulsion of iodoform in glycerin injected. The operation was performed under strict antiseptic precautions, and the puncture was sealed with iodoform collodion. After the completion of the operation the pulse and respirations were good and there was apparently little if any shock, but the urine was very scanty, and within twelve hours grave symptoms of iodoform-intoxication appeared, to which the patient succumbed at the end of forty-eight hours. The writer has witnessed mild symptoms of iodoform-intoxication, in a number of cases of renal affections that required an operation, when only a very small quantity of the drug was used, and at present limits its use or dispenses with it entirely in operations upon patients suffering from renal disease. During and after operations for renal tuberculosis the patient's body should be well protected, and the external artificial heat should be continued until he has fully rallied from the immediate effects of the operation. Renal insufficiency should receive prompt attention, and, if stimulation is required, coffee, ammonia, and strychnine should be administered either subcutaneously, by the stomach, or by the rectum. If symptoms of shock continue for some time after the operation, a physiological solution of salt should be administered by intravenous or subcutaneous injection. From 12 to 16 ounces of this solution will often be followed by prompt effects upon the action of the heart and will result in increasing the action of the remaining kidney.

Nephrotomy.—Incision and drainage of the kidney through the lumbar region are called for in all cases of tubercular pyonephrosis in which a primary nephrectomy is contraindicated. It is an exceedingly valuable palliative measure, as it relieves pain by diminishing tension and furnishes a direct and short outlet to the liquefied tubercular material that has accumulated in the pelvis or within the capsule of the kidney, and by doing so cuts off the irrigation of the urinary tract with tubercular pus or urine. Occasionally the operation proves curative, as it secures free access to the seat of the disease, which can then be treated by antibacillary solutions. Catheterization of the ureter through the fistulous opening is possible in many cases, and in this way the affected portion of the ureter can be reached and subjected to direct medication. Adams,[1] of Chicago, treated a case of tubercular pyelo-ureteritis in this manner, and finally effected a complete cure. Tuffier[2] regards retention of pus and pain as the most important indications for operative treatment. He lays down the rule that if the affection is unilateral, nephrectomy is indicated, and in doubtful cases nephrotomy. He made 25 nephrotomies with a mortality of 47.8 per cent., and only 2 definitive cures; 57 nephrectomies, of which 11 were made through the abdomen, with a mortality of 36.3 per cent., and 46 through the lumbar region, with a mortality of 28.2 per cent.

Newman gives the following table showing the mortality that attends nephrotomy in cases of tubercular pyonephrosis:

	Males.	Females.	Recoveries.	Deaths.
Lumbar	7	12	13	6 = 32 per cent.
Abdominal	1	..	1	—
	8	12	14	6 = 30 per cent.

[1] Personal communication.
[2] "Étude Anatomo-pathologique et Clinique sur la Tuberculose rénale." *Arch. Gén. de Méd.*, May, June, 1892.

The same author tabulates nephrectomy cases for the same affection as follows:

	Males.	Females.	Recoveries.	Deaths.
Lumbar	13	13	16	10 = 38.5 per cent
Abdominal	..	7	5	2 = 28.6 per cent.
	13	20	21	12 = 36.3 per cent.

The kidney may be exposed very satisfactorily by Simon's vertical incision along the outer border of the sacrolumbalis, extending from the lower margin of the last rib to the crest of the ilium. After incising the skin and fascia the kidney should be exposed by the use of blunt instruments, using the knife as sparingly as possible. The adipose capsule can be torn and the margins of the rent retracted. On exposure of the fibrous capsule of the kidney, an exploratory puncture may be made if any doubt remains concerning the diagnosis. In operations for tuberculosis of the kidney by lumbar incision and drainage, it is always necessary to make the visceral incision into the pelvis of the organ. If the kidney is not much enlarged, this is not always easy without a guide. The pelvis of the kidney can always be reached and accurately located by the needle of the exploring-syringe. In the absence of enlargement of the kidney and fluctuation, it is advisable to locate the pelvis of the kidney in this manner, and use the needle as a guide in making the visceral incision. If the cortex of the kidney has not been diminished in thickness by pressure or destroyed by the disease, the incision should be made with the knife-point of the Paquelin cautery, as the use of the knife in performing this part of the operation is often attended with free and even dangerous hemorrhage. To obviate this source of danger the platinum point should be heated only to a dull-red heat and advanced very slowly. The entrance of the cautery into the tubercular focus or into the pelvis of the kidney is announced by the escape of urine or tubercular material. The opening into the pelvis of the kidney may be enlarged sufficiently by inserting the index finger or by the use of large hemostatic forceps. The opening should always be large enough to enable the operator to explore the tubercular cavity with the index finger. A rubber drain the size of the incision is now inserted into the tubercular cavity, and secured by a large safety-pin. The wound should not be sutured, but the space around the tube is packed with strips of iodoform or plain sterile gauze.

The condition of the urine after nephrotomy is a valuable indication concerning the state of the opposite organ. If the urine clears up and remains in a normal condition, it is an indication that the opposite kidney is intact. If the urine under the microscope presents the same evidences of tuberculosis as before the operation, it is safe to assume that the disease has reached the bladder or that the opposite organ is involved. If the disease is bilateral, subsequent radical measures are absolutely contraindicated. In such cases it may become necessary to repeat the operation on the opposite side, and subject the kidney to direct treatment by irrigating the tubercular cavity daily with an aqueous solution of iodine or trichloride of iodine. Balsam of Peru is a valuable remedy in the treatment of all suppurative tubercular affections, and can be used to advantage in these cases. If suppuration is profuse, injections of peroxide of hydrogen are useful as an antiseptic. If the subsequent condition of the urine furnishes satisfactory evidence that the opposite kidney is intact, a secondary nephrectomy should be performed, in the absence of other contraindication and in case the local treatment through the fistula does not prove satisfactory. A nephrotomy is a valuable preparatory measure, under such circumstances, to a secondary nephrectomy. If the kidney operated upon continues to secrete a considerable quantity of urine, a rubber receptacle holding a pint should be connected with the drain by a short piece of rubber tubing and held in place by a bandage. A patient operated upon by the writer ten years ago for hydronephrosis has been wearing this simple apparatus with comfort ever since.

Nephrectomy.—Primary nephrectomy is the operation of choice in the treatment of unilateral renal tuberculosis, provided the patient is not suffering at the same time from advanced pulmonary phthisis or from tuberculosis of any other important organ beyond the reach of successful surgical treatment. Secondary nephrectomy may become necessary if, after a nephrotomy, all doubts regarding the healthy condition of the opposite kidney have been set aside. The greater difficulties encountered in the diagnosis of renal tuberculosis in men than in the diagnosis of the same condition in women accounts for the greater frequency with which primary nephrectomy has been performed upon women than upon men. An additional reason is the fact that in men the kidneys are more frequently involved by an ascending tubercular process which, as a rule, attacks both kidneys at the same time or after an interval of variable duration. The opposite kidney was normal in 12 out of 24 cases of renal tuberculosis examined by Steinthal,[1] while in those cases in which it was affected the disease had often progressed only to a limited extent.

The curative effect of the operation and the mortality attending it are variously estimated by different authors. The comparatively great mortality is undoubtedly due to the frequency with which operations have been performed upon patients suffering from bilateral disease. Even if the opposite organ is intact, it may not assume the necessary functional activity to compensate for the loss of the diseased organ. Danger from this source is slight if the parenchyma of the organ removed has been entirely destroyed. In two of the writer's cases operated upon for tuberculosis of the kidney, death resulted from uremia in one case in two days after the operation. Both patients were females, thirty and thirty-five years of age. In one case the disease had existed for two years and in the other for three years. In both cases the infection had its starting-point in the pelvis of the kidney. In the case in which death occurred on the second day the opposite organ was very much enlarged and extremely vascular, but there was no trace of tuberculosis. The kidney removed contained very little glandular tissue. The kidney and pelvis were transformed into a sacculated cavity holding about four ounces, filled with

[1] *Virchow's Archiv*, B. cxxv, p. 25.

liquefied cheesy material. No tuberculosis existed in any other organ. In the second case the kidney removed was not much enlarged; its pelvis contained about a tablespoonful of tubercular pus. The cortex was infiltrated with tubercles some of which had undergone caseation. This patient was very anemic, but was not much emaciated. The healthy condition of the opposite kidney was demonstrated by a clinical and microscopical examination of the urine drawn from the pelvis with a ureteral catheter. The post-mortem showed that the opposite kidney was very slightly enlarged and very vascular. No tubercles were found. The apex of the left lung was the seat of an old tubercular process which evidently had been in a latent condition for many years. In the first case not a drop of urine was secreted after the operation; in the second case about four ounces were withdrawn from the bladder a few hours after the operation, after which the suppression was complete, and death from uremic intoxication occurred four days later. In two cases an apparent permanent cure followed the operation, as the patients remain in good health two and four years after the operation. In one case the opposite organ became affected by an ascending tubercular process from the bladder six months after the operation, followed by death three months later.

Rosenstein states that he is acquainted with two cases of nephrectomy for unilateral tuberculosis on patients remaining in good health even after years. Facklam[1] gives 40.9 per cent. as the proportion of

---

[1] *Arch. f. klin. Chir.*, B. 1. Heft 4.

permanent cures. He found in the literature and by private inquiry 88 cases of nephrectomy for tuberculosis, of which 25 died from the immediate effects of the operation.

Newman[1] has collected 33 cases of nephrectomy for tuberculosis; in 26 cases the lumbar operation was performed—13 males with 5 deaths, and 13 females with 5 deaths—and in 7 females the abdominal operation was adopted, with 2 deaths. The causes of death in the 12 fatal cases were as follows: Shock and collapse, 4; septicemia, 1; uremia, 1; hemorrhage, 2; vomiting, 1; cause unknown, 3. In many of these cases the patient recovered from the operation and lived for a few months, but in how many instances a cure, in the proper sense of the term, was effected he was unable to determine.

The first intentional, well-planned nephrectomy was performed by Simon[2] on August 2, 1869. As the case is of great historical and scientific interest, an abstract of the full report will be given.

The patient was a woman forty-six years of age, who, a year and a half before, had been operated upon by Walther for an ovarian tumor. Firm adhesions between the tumor and the uterus made it necessary to remove the whole body of the uterus; the opposite ovary was also removed. During the operation the left ureter was injured, and a urinary fistula remained at a point corresponding with the lower angle of the wound. As failure attended various methods of treatment that were resorted to to close the fistula, and later to destroy the secreting

[1] *Op. cit.*   [2] *Chirurgie der Nieren.* Erlangen, 1871.

structure of the kidney by closing the proximal end of the ureter, he decided upon removal of the kidney as offering the only chance of freeing the patient from the great annoyance and discomfort caused by the continuous escape of half the urine through the fistula. He made ample preparations by performing the operation repeatedly on the cadaver in men to develop the necessary technique, and by removing one of the kidneys in dogs to determine the safety of the procedure. Having planned the operation and ascertained its comparative safety, he removed the kidney through a vertical incision extending from the eleventh rib to near the crest of the ilium, following the outer border of the sacro-lumbalis muscle. The adipose capsule was incised, and after detaching it from the kidney the pedicle was transfixed and tied. A small portion of the kidney-tissue remained at the end of the stump, but outside of the ligature. Three bleeding points after cutting through the pedicle were ligated separately. No untoward symptoms referable to the opposite kidney developed, which satisfied the operator that the opposite kidney was assuming in a satisfactory manner the necessary compensatory functions. The recovery of the patient was retarded by an attack of erysipelas. The patient regained her former health and buoyancy of spirits, and did excellent service as a nurse during the Franco-Prussian war.

The operation as practised by Simon has been modified by different surgeons, as it was found that Simon's incision did not afford sufficient room for the removal of a diseased kidney, especially in cases in which the renal affection is complicated by firm adhesions to important organs or by extension of the tubercular process to the paranephric tissues. So many cases of absence of one kidney have been reported that it becomes a duty on the part of the surgeon, before he decides to make a nephrectomy, to ascertain such a possible anatomical defect by the use of the cystoscope if the organ cannot be felt by palpation. The cystoscope in such cases will demonstrate the absence of the ureteral orifice, and will show that all the urine escapes from one ureter. In doubtful cases it is good practice to make a lumbar exploratory incision down to the supposed healthy kidney. However, in the majority of the cases of unilateral renal disease the healthy kidney is so much enlarged that it can readily be palpated. The writer has recently had a very interesting experience in which nephrotomy performed for hydronephrosis demonstrated the fact that the patient had only one kidney. The patient was a man twenty-five years of age, who for years had suffered intensely from what was diagnosticated as intermittent hydronephrosis. Attacks of excruciating pain in the region of the left kidney set in every few days, followed by complete relief during the intervals. For several months he noticed that during these attacks a swelling appeared in the region of the kidney and disappeared quite suddenly after the cessation of the pain. During the attacks no urine was voided, while a copious discharge of urine always followed each attack. The writer saw him during several of these attacks, and had no difficulty in outlining a fluctuating swelling in the region of the left kidney. There was no doubt as to the existence of hydronephrosis. The urine was normal, and, according to the patient's statements, was stained with blood only on two or three occasions. He was under medical treatment for a long time without being benefited, when he was sent by his attending physician to the surgical clinic of Rush Medical College for operative treatment. The attacks came on every three to six days and lasted for two or three days. With the beginning of pain the urine became suppressed, and remained so during the whole attack, followed by the appearance of the renal swelling. At the end of the attack from one to two quarts of clear normal urine would be voided. That the bladder contained no urine during the attacks was demonstrated on several occasions by the use of the catheter. Toward the end of each attack the patient showed plain evidences of uremic intoxication—intense headache and constant retching and vomiting. The operation was performed during one of these attacks, no urine having escaped for seventy-two hours, and the patient being semi-comatose from uremic intoxication. No anesthetic was used. Local anesthesia by Schleich's infiltration-method was employed, and did good service in diminishing the pain. The kidney was exposed by Simon's vertical incision. The adipose capsule was normal. On direct palpation of the kidney distinct fluctuation could be elicited, but the sac felt firm and resistant. The kidney was incised with the knife-point of the Paquelin cautery. What attracted the attention of all who witnessed the operation was the thickness of the parenchyma of the kidney before the distended pelvis was reached.

About two quarts of clear urine escaped. Although the hydronephrotic sac was nearly as large as an adult's head, examination showed kidney-substance over an inch in thickness throughout. The writer could not account for this enormous hypertrophy of the organ. Digital exploration of the interior of the sac revealed the characteristic conditions of hydronephrosis. In searching for the ureteral orifice a stone of very peculiar formation was found impacted on one side in the ureter and projecting on the other side into the bladder. The stone was smooth, hard, about an inch in length, and tapered on the ureteral side in such a way that it fitted into the ureter like a cork in a bottle. The course of the intermittent hydronephrosis had been found and removed. The sac was drained and the usual antiseptic dressing applied. A very small quantity of urine escaped from the bladder soon after the operation, after which every drop of urine escaped through the fistula. The ureter was catheterized from the pelvis of the kidney repeatedly, and as it was strictured at a point a little below where the stone was lodged, the ureter was dilated. The fistulous opening remained for over three months; after that time some of the urine was passed *per vias naturales*, and in the course of a few weeks the fistula closed permanently. The patient regained his former health, and remains well nearly a year after the operation. It is plain that in this case there was no kidney on the other side, which accounts for the enormous hypertrophy of the one operated upon, and the complete retention of urine when the stone became impacted in the ureter. This and many similar cases should make the surgeon cautious when he decides upon the propriety of performing a nephrectomy for renal tuberculosis, as it may be possible that the diseased kidney is the only one, as in a case reported by Lange of New York, the one stated above, and a few others.

Technique of Operation.—Most operators at the present time favor the lumbar operation, as statistics show that the extraperitoneal route, on the whole, is the safest. The patient should be placed on the opposite side, with a firm round roll under the lumbar region for the purpose of enlarging the space between the last rib and the crest of the ilium. This space is often extremely narrow in rachitic subjects and in women who have been in the habit of lacing tightly. Resection of the last rib is now seldom performed to enlarge this space, but might become necessary in the removal of a kidney that has not descended below the normal level or is attached to important organs, its safe isolation requiring an abundance of space. The incisions selected at the present time by different surgeons to reach the kidney through the lumbar region are—1. Vertical; 2. Oblique; 3. Transverse. The vertical or Simon's incision extends from the last rib to the crest of the ilium, along the outer border of the sacrolumbalis muscle. The manner of making the incision has been explained under the discussion of nephrotomy. This incision is now seldom employed when it is the intention to remove the kidney. König's incision is a prolongation of Simon's incision by carrying it to a point a little above the iliac crest, obliquely inward and downward to near the spine of the ilium, and then along the upper border of the outer third of Poupart's ligament. This incision affords ample room in the majority of cases, and enables the operator to follow extraperitoneally the ureter to near the bladder, and, if found necessary, to remove it with the kidney. The incision through

FIG. 26.—Exposure of kidney by transverse incision.

the muscular layers can be made the whole length of the vertical incision, when the adipose capsule is exposed, and the muscular layers from here can be cut between the outer and middle fingers of the left hand, with which the parietal peritoneum is pushed aside. The transverse incision (Fig. 26) is commenced a little below the last rib, at the outer border of the sacrolumbalis muscle, and is carried in a transverse direction as far as may become necessary. The muscular layers are divided in a similar manner from the adipose capsule, after detaching the peritoneum with the index and middle fingers. Should the peritoneal cavity be opened at any time, the rent must be sutured at once before proceeding further with the operation.

The first step in the operation consists in exposing the kidney as freely as possible. All hemorrhage is arrested promptly with the aid of hemostatic forceps. Before the kidney itself is attacked, the field of operation must be made absolutely bloodless, to facilitate the most important part of the operation, the isolation and removal of the kidney. Bardenheuer devised an incision with the sole intention of affording an abundance of room. He makes Simon's vertical incision, and joins this with two transverse incisions toward the rectus muscle—one at its upper angle, parallel with the last rib, and the other at its lower angle, just above and parallel with the crest of the ilium. By raising a quadrangular flap in the manner of a trap-door, the affected organ, if enlarged and below its normal level, is freely exposed. This trap-door incision necessitates much more traumatism than either the transverse or the oblique incision, without affording sufficient advantages to offset the increased risks incident to the operation, and is therefore seldom resorted to at the present time. James Israel, who has had such a large experience in renal surgery, invariably makes a transverse incision, and claims that he has never had to change his plans for the removal of kidneys the seat of large tumors or inflammatory conditions of the most complicated kind. As soon as the adipose capsule of the kidney has been freely exposed by any of the incisions that have been mentioned, another examination of the kidney and its surroundings should be made. If the disease has extended to the capsule, its removal with knife and scissors or with the sharp spoon becomes necessary. If the disease is complicated by a paranephric abscess, this should be curetted thoroughly, and after washing out and drying the cavity the kidney is removed in the usual manner. In such cases it is usually better to perform the operation in two stages, postponing the nephrectomy until the cavity is lined with healthy granulations and the patient has recovered from the effects of the first operation. If this is the plan adopted, the cavity should be packed with plain sterile gauze, interposing between the gauze packing and the inner surface of the cavity one or two layers of iodoform gauze. The external wound must be kept open widely until the second operation is performed.

If the adipose capsule of the kidney has become transformed into a dense fibrous mass in which the tubercular kidney is imbedded firmly and immovably, its capsule intimately connected with the fibrous mass, the removal of the whole mass, inclusive of the kidney, is a very difficult and most dangerous operation, as the fibrous mass often includes not only the kidney but also its vessels, and not infrequently the vena cava. Confronted by such difficulties, the surgeon is often glad to avail himself of Ollier's subcapsular nephrectomy, which minimizes the danger, without interfering with the removal of all the infected tissue.

If the adipose capsule is normal—that is, if the disease remains limited to the kidney—an ideal nephrectomy is the operation of choice. The adipose capsule is laid open by an incision parallel with the external cut, after which, with the finger-tips and blunt instruments, the kidney is laid bare down to its hilus. The use of sharp instruments in this part of the dissection is limited to the cutting of fibrous strings or bands, but, as a rule, they can be dispensed with entirely. The next step in the operation consists in delivering the kidney through the incision. This is done by bringing forward first the upper part of the organ. Harmful traction upon the renal vessels must be carefully avoided. If the renal vessels are short, altered in their structure by disease, or incorporated in a mass of scar-tissue, they might be torn during this step of the operation. In one of Luecke's cases of laparonephrectomy this accident occurred. The renal vein was torn off at its junction with the vena cava, and the fearful hemorrhage that followed, although finally controlled, was the immediate cause of death, which occurred soon after the operation.

It is not always possible, on reaching the pedicle, to ligate separately the ureter and the renal vessels. Good retractors and reliable assistants are necessary in performing this part of the operation. It must also be remembered that the vascular supply of the kidney is not infrequently atypical. The kidney may be supplied with blood from two or more renal arteries, and there may be more than one renal vein. The vessels also vary greatly in length.

The possibility of tuberculosis of a horseshoe kidney should also be borne in mind by the operator in performing nephrectomy for tuberculosis. If such a case should present itself, it would probably be better to abandon the operation, as the removal of one half of a horseshoe kidney for tuberculosis would be an extremely uncertain and dangerous operation.

In removing the kidney for tuberculosis it is very important to remove the entire pelvis and as much of the tubercular ureter as is within reach through an oblique extraperitoneal incision. It has happened that after the extirpation of the kidney for unilateral disease the symptoms for which the operation was performed remained, and the surgeon was later compelled to remove the tubercular ureter. Wherever it can be done, the ureter should be exposed and carefully examined some distance from the pelvis of the kidney before the pedicle is tied. The isolation of the pedicle should be done, as far as is practicable, by the use of a blunt instrument. If the ureter is affected, it should be tied at the lowest possible level, after which the entire pedicle is tied *en masse*. After the removal of the kidney the lumen of the renal artery can be found without much difficulty, and a separate ligament is applied to this vessel near its cut end. It is advisable to grasp the pedicle with a pair of toothed hemostatic forceps before the kidney is detached, so that in the event of hemorrhage the stump may be drawn forward and the bleeding vessels tied separately. If this precaution is neglected, the retracted stump is often difficult to find and the hemorrhage incident to such an accident is difficult to control. If the ureter is intact after the ligation of the pedicle, it, as well as the larger renal vessels, is tied separately upon the surface of the stump.

Nephrectomy for inflammatory disease is a much more difficult operation than for tumors, as the process in the latter instance in operable cases is limited by the capsule, while pararenal affections very frequently complicate renal tuberculosis. Different instruments of special construction for the ligation of the pedicle have been invented, notably the "bill forceps" by Lange. Such instruments may answer a very useful purpose in special cases, but they can be dispensed with by substituting for them hemostatic forceps with different curves and of different sizes, used in other especially pelvic and abdominal operations. With a well-curved Péan forceps a ligature can be thrown around a short and thick pedicle. Slipping of the ligature *en masse* must be guarded against by cutting the pedicle, especially its central portion, sufficiently far away from the line of constriction. If it is found impossible to ligate the pedicle—and such cases will occur—the hemorrhage should be prevented and treated by the use of forcipressure-forceps, which are incorporated in the dressings and removed at the end of forty-eight or seventy-two hours. Slipping of the forceps must be prevented during this time by applying over the absorbent and elastic dressing a few turns of a plaster-of-Paris bandage.

The writer can imagine cases in which the kidney

cannot be wholly removed, owing to extensive destruction of its substance and great thickening around it. It is in such cases that the surgeon has to remove it piecemeal—that is, by *morcellement.* The cavity after removal of the kidney by this method should be thoroughly cauterized with the Paquelin cautery and treated by tamponade, plain aseptic gauze impregnated with balsam of Peru being used for this purpose. There are also cases in which great debility militates against completing the operation.

Lumbar nephrectomy in two stages has a field of usefulness, as by dividing the operation into two stages the shock would necessarily be favorably modified. Shock following a complete nephrectomy figures largely in the mortality statistics. The first step of the operation, exposing the kidney, can be performed without an anesthetic by resorting to Schleich's infiltration-method. After the incision has been made the wound should be kept well open by packing with simple sterile gauze. In the course of a few days the organ can be removed with the patient under the influence of a general anesthetic. The second operation can be performed without the use of the knife. The writer is satisfied that by performing nephrectomy in two sittings, in cases requiring conservation of strength and blood, patients can be saved that would succumb to the immediate effects of a complete nephrectomy.

Subcapsular Nephrectomy.—The removal of the kidney without its capsule—*néphrectomie sous capsulaire* of Ollier—becomes an operation of necessity in certain cases of renal tuberculosis in which, owing to firm adhesions of the kidney to the surrounding tissues, a typical nephrectomy cannot be performed. In both forms of renal tuberculosis the organ is often not much enlarged, and frequently is fixed in a mass of firm connective tissue which largely takes the place of the adipose capsule. This condition often renders it impossible to remove the capsule, when the surgeon is obliged to enucleate the kidney and leave its capsule behind.

Partial Nephrectomy.—Experiments on animals made by Tuffier and others have demonstrated the reparative capacity of the kidney. It has been shown that a considerable portion of the organ can be removed with prompt healing of the visceral wound, followed by partial, if not complete, regeneration of the lost tissues. The results of these experiments have led to the performance of partial nephrectomy in cases in which the injury or disease is limited. James Israel[1] was the first one to resort to partial nephrectomy for localized renal tuberculosis. He removed successfully one-half of a tubercular kidney. The patient was a woman twenty-three years old, who three months before her admission into the hospital had suffered from vesical colic, frequent urination, and renal colic that appeared every two or three days. The left kidney, the seat of the disease, could not be felt by palpation. The urine contained blood-corpuscles, but no bacilli. After enucleation of the kidney from the adipose capsule it was found that only the upper half of the organ was affected; this portion was resected, the

---

[1] *Centralb. f. Chir.*, 1896, No. 13.

hemorrhage being controlled by digital compression of the renal vessels. The surface bleeding, which at first was quite profuse, yielded to a gauze compress fastened in place by suturing. In the portion of the kidney that was left small tubercles were seen between the cortex and the medullary portion. The patient recovered, and was in good health and five months pregnant a year after the operation. Although this case terminated favorably, partial nephrectomy for infective diseases of the kidney has necessarily an exceedingly limited field of application, especially in tubercular affections of this organ.

Laparonephrectomy.—The removal of a tubercular kidney through an abdominal incision is attended by greater immediate and remote risks to life than the operation of lumbar nephrectomy. The increased shock and the liability of infecting the abdominal cavity are sources of danger that militate against this method of operating. On the other hand, it has been stated that the presence of the opposite kidney and its condition can be ascertained by direct palpation, and that the pedicle can be reached with more ease and certainty than through a lumbar incision. Statistics from different sources, with few exceptions, are in favor of the lumbar operation. More thorough preparation of the patient is necessary if the operator selects the abdominal route. The minutest precautions must be carried out to prepare the patient for the abdominal operation. At least a week should be spent in carrying out the necessary antiseptic and aseptic precautions. The incision through the abdominal wall is made directly over the swelling, from the costal arch downward sufficiently to give free access to the kidney. In the absence of a palpable swelling the incision is made through the linea semilunaris, as advised by Langenbuch. As soon as the kidney has been located the intestines are pushed to the opposite side and protected with a large flat sponge or a compress of sterile gauze wrung out of a warm physiological solution of salt. The parietal peritoneum overlying the kidney is then incised the whole length of the organ in a vertical direction. With fingers and blunt instruments the kidney is freed from its adipose capsule, when the pedicle is reached and secured in the manner described in connection with lumbar nephrectomy. Drainage of the cavity through an opening in the lumbar region is indicated in the majority of cases. This can readily be done by pushing, at a point corresponding with the lowest part of the cavity, a strong pair of hemostatic forceps from within outward through the muscles, and cutting a small opening in the skin directly over the point raised by the forceps. A rubber drain the size of the little finger should now be drawn through from without inward, and cut about an inch above the level of the posterior wall. The peritoneal wound should be closed carefully, either fine catgut or silk sutures being used. The abdominal wound is then closed in the usual manner, and protected with an antiseptic absorbent dressing held in place with long, broad, adhesive strips and an abdominal bandage. Some surgeons prefer to stitch the margins of the posterior peritoneal incision to the external wound before attempting the removal of the kidney, in order to guard the peritoneal cavity more securely against infection.

The writer can see how abdominal nephrectomy could be deprived of the greatest sources of danger to life by performing the operation in two stages. The first part of the operation would consist in creating free access to the kidney by suturing the margin of the posterior peritoneal incision to the peritoneum of the abdominal incision, packing the cavity with sterile gauze, and closing the remaining portion of the incision by temporary sutures. In the course of a few days, after the kidney has been rendered, to all intents and purposes, an extraperitoneal organ, the second operation could be performed in the manner described, by removing the stitches and reopening the wound. This modification of laparo-nephrectomy certainly deserves a trial in some of the cases in which it is more advantageous to remove the kidney by the transperitoneal method. During the operation by the abdominal route shock should be reduced as much as possible by securing the necessary degree of external artificial heat and by the subcutaneous administration of strychnine. Saline infusions by one of the different routes are indicated if the patient does not rally promptly from the immediate effects of shock. The function of the opposite kidney must be carefully watched, and any failure to assume compensatory action should be met by appropriate treatment.

INDEX.

ABSCESS of spine, tubercular, rupture of, into pelvis of the kidney (Fig. 24), 238

BACILLI, staining of, in tubercular urine, 270
Bacillus tuberculosis, specific effect of, 93
Bladder, chronic inflammation of, histological elements found in the urine in, 194
   tuberculosis of, 185
      complications, 189
      diagnosis, 191
         from renal tuberculosis, 199
      etiology, 189
      pathology, 186
      prognosis, 200
      symptoms, 191
      treatment, 201
         cystotomy, perineal, 208
         suprapubic, 211
         general, 203
         medication, direct, 205

CARCINOMA complicating tubercular endometritis, 145
Castration, double, in tuberculosis of testicle, 77
   in tuberculosis of the testicle, 74
   results of, 75

Catheter, Senn's self-retaining, for suprapubic drainage, 211
Cauterization in tuberculosis of testicle, 79
Chancre, urethral, differential diagnosis of, 31
Coitus, inoculation with tuberculosis during, 19
Creosote in treatment of tuberculosis of female generative organs, 98
Curettage in tuberculosis of epididymis, 79
Curettement in treatment of tuberculosis of the penis, 23
Cystoscopy in diagnosis of renal tuberculosis, 277
Cystotomy, perineal and suprapubic, in treatment of vesical tuberculosis, 208, 211
Cysto-uretero-pyelonephritis tubercular (Fig. 16), 186
Cysts, ovarian, tubercular, 177

DRAINAGE, suprapubic, Senn's self-retaining catheter for, 211

ENDOMETRITIS, tubercular, complicated by carcinoma, 145
   in child-bearing women, 143
Epididymis, tuberculosis of, 47
   diagnosis, 67
   etiology, 49

Epididymis, tuberculosis of, frequency, 58
  manner of infection, 48
  pathology, 55
  predisposing causes, 51
  symptoms, 67
  treatment, 73
Epididymitis, chronic, differential diagnosis of, 70
  gonorrheal, as a precursor of tuberculosis of the epididymis, 54
  tubercular, 47
    castration for, 74
Esthiomene, 103
Evidement in tuberculosis of epididymis, 79

FALLOPIAN TUBES, frequency of affection of, in genital tuberculosis, 85
  tuberculosis of, 149
    diagnosis, 164
    etiology, 160
    frequency, 150
    pathology, 152
    predisposing causes, 164
    treatment, 167
      operative, 169

GENERATIVE ORGANS, female, tuberculosis of, 82
  diagnosis, 95
  frequency, 91
  manner of infection, 84, 90
  pathology, 93
  prognosis, 97
  treatment, 98
Genital organs, male, tuberculosis of, 17
Granulations composées, 65

Guaiacol in treatment of tuberculosis of female generative organs, 98

HEMATURIA as a symptom in tuberculosis of the bladder, 196

INOCULATION - EXPERIMENTS in diagnosis of renal tuberculosis, 277
Insufflation, rectal, in diagnosis of renal swellings, 274
Iodoform-glycerin injections in treatment of tubercular abscesses, 100
Iodoform injections in treatment of tuberculosis of female generative organs, 100
  Senn's syringe for, 99
  intoxication from, in treatment of renal tuberculosis, 287

KIDNEY, absence of one, 297
  exposure of, by Simon's vertical incision, 290
    by transverse incision, 301
  horseshoe, tuberculosis of, as a factor in performing a nephrectomy, 305
  infection of, with tubercle bacilli, 226
  palpation of, 272
  percussion of, 274
  tubercular, removal of, through an abdominal incision, 309
  tuberculosis of, 222
    classification, 231
    diagnosis, 259
    diagnostic aids, 227
    etiology, 246
    frequency, 226
    operations for, 284

Kidney, tuberculosis of, pathology, 231
　precautions to be observed in operations for, 286
　predisposing causes, 255
　prognosis, 278
　progressive nature of, 279
　symptoms, 259
　treatment, 280
　　general, 281
　　operative, 284
　varieties: caseous, 235
　　miliary, 233
　　tubercular, 240
König's incision in nephrectomy, 300

LAPARONEPHRECTOMY, 309
　in two stages, advantages of, 311
Lupus of vulva, 104
Lymphoid tubercle, 64

MEATUS, secondary tubercular ulcer of, 31
Microscope, use of, in diagnosis of tuberculosis of female generative organs, 95
Morcellement, treatment after, 307

NEPHRECTOMY, 292
　abdominal, 309
　　in two stages, advantages of, 311
　first operation of, 295
　König's incision for, 300
　lumbar, in two stages, 307
　mortality of, 290, 293
　partial, 308
　Simon's operation, 296
　statistics of, 285, 290, 295
　subcapsular, 307
　technique of operation, 300
　transverse incision for, 301

Nephritis, caseous, 235
　chronic (Fig. 23), 236
　miliary, 233
　tubercular, 240
Nephro-phthisis, 236, 238
Nephro-pyelo-ureteritis (Fig. 19), 217
Nephrotomy, 288
　mortality of, 289
　statistics of, 285, 289

ORCHITIS, chronic, differential diagnosis of, 70
　tubercular, 56
Ovary, tuberculosis of, 173
　diagnosis, 182
　etiology, 180
　pathology, 175
　symptoms, 182
　treatment, 183

PALPATION, rectal, in diagnosis of renal swellings, 275
　renal, 272
Penis, extensive destruction of, from a tubercular process, 20
　tuberculosis of, 19
　curettement in, 23
Percussion, renal, 274
Peri-oöphoritis, tubercular, 176
Peru, balsam of, in treatment of suppurative tubercular affections, 291
Phthisis, renal, 238
Prostate, tuberculosis of, 40
　symptoms, 45
　treatment, 46
Puerperium, frequency of uterine tuberculosis after, 135
Pyelonephritis, tubercular, 240
Pyonephrosis, tubercular, 240

SALPINGITIS, tubercular, 149
"Scrofula" of the testicle, 68

Scrofulous testicle, conservative operation for, 77
Seminal vesicles, eradication of, in tuberculosis, 36
tuberculosis of, 34
Senn's self-retaining catheter for suprapubic drainage, 211
syringe for iodoform injections, 99
Simon's operation of nephrectomy, 296
Spermatic cord, removal of, in tubercular spermatitis, 34
tuberculosis of, 33
Spermatitis, tubercular, 34
Spine, tubercular abscess of, rupture of, into pelvis of kidney (Fig. 24), 238
Staining of bacilli in tubercular urine, 270
Stricture of urethra, tubercular, 32

TESTICLE and epididymis, tuberculosis of, 47
diagnosis, 67
differential, 73
general dissemination of, 65
symptoms, 67
treatment, 73
Testicle, chronic inflammation of, 55
" scrofula " of, 68
tuberculosis of, castration in, 74, 77
development of, 61
double, 53
microscopic appearance of tubercle-tissue, 60
primary chronic, 57
treatment, 73
Testis, scrofulous inflammation of, 55

Thermometer, use of, in diagnosis of tuberculosis of female generative organs, 96
Tubercle bacilli, elimination of, through the kidneys, 247
lymphoid, 64
Tubercles, ovarian, histological structure of, 176
unusual size of, in tuberculosis of the testicle, 62
Tuberculosis, bacillus of, specific effect of, 93
genito-urinary, origin and dissemination of, 224
inoculation with, during coitus, 19
in the negro, rapid course of, 21
intra-uterine, 140
miliary, 233
pulmonary and genital, etiological relation, 223
pulmonary, preceding genital tuberculosis, 86
renal, artificial production of, 251
transmission of, from mother to fetus, 138, 141

ULCER of meatus, secondary tubercular, 31
Ulceration of urethra, tubercular, in pulmonary tuberculosis, 31
Ureter, tuberculosis of, 216
pathology, 219
symptoms, 220
treatment, 220
Ureters, catheterization of, as a diagnostic aid in diagnosis of tuberculosis of the kidney, 227
in diagnosis of renal tuberculosis in women, 275
changes in, in tuberculosis of the kidney, 231

Urethra, chancre of, differential
  diagnosis, 31
 tubercular stricture of, 32
  ulceration of, in pulmonary tuberculosis, 31
 tuberculosis of, 28
  in women, 32
  manner of infection, 28
  treatment, 33
Urethritis, tubercular, 30
Urinary system, tuberculosis of,
 chronological order of infection
 of different parts, 239
Urine, bacteriological examination
  of, in diagnosis of tuberculosis of the kidney, 228–
  230
 condition of, after nephrotomy,
  291
 examination of, for bacilli, in
  diagnosis of tuberculosis
  of the bladder, 197
  in diagnosis of renal
   tuberculosis, 267
 histological elements found
  in, in chronic inflammation
  of the bladder (Fig. 17),
  194
 incontinence of, in tubercular
  urethritis, 30
 of tubercular patients, infectiveness of, 249

Urine, retention and incontinence
  of, in tuberculosis of the
  bladder, 196, 197
 sedimentation of, means of
  hastening, 269
 tubercular, staining of bacilli
  in, 270
Uterus, tubercular, section from
  the fundus of (Fig. 10), 128
 tuberculosis of, 122
  diagnosis, 143
  etiology, 133
  frequency, 122
  pathology, 126
  prognosis, 146
  symptoms, 143
  treatment, 146

VAGINA, tuberculosis of, 111
 clinical cases, 112
 diagnosis, 119
 etiology, 114
 pathology, 114
 predisposing causes, 119
 treatment, 120
Vesiculitis, tubercular, 34
Vulva, lupus of, 104
 tuberculosis of, 102
  diagnosis, 108
  pathology, 107
  treatment, 109

Standard Medical and Surgical Works

PUBLISHED BY

W. B. Saunders, 925 Walnut Street, Philadelphia.

MR. SAUNDERS, in presenting to the profession the following list of publications, begs to state that the aim has been to make them worthy of the confidence of medical book-buyers by the high standard of authorship and by the excellence of typography, paper, printing, and binding.

The works indicated in the Index (see next page) with an asterisk (*) are sold by subscription (*not by booksellers*), usually through travelling solicitors, but they can be obtained *direct* from the office of publication (charges of shipment prepaid) by remitting the quoted prices. Full *descriptive circulars* of such works will be sent to any address upon application.

All the other books advertised in this catalogue are commonly for sale by *booksellers* in all parts of the United States; but any book will be sent by the publisher to any address (post-paid) on receipt of the price herein given.

CONTENTS.

Anatomy.
	PAGE
Haynes, Manual of Anatomy	24
Nancrede, Anatomy and Manual of Dissection	16
Nancrede, Essentials of Anatomy	26

Bacteriology.
Ball, Essentials of Bacteriology	26
Crookshank, A Text-Book of Bacteriology	13
Frothingham, Laboratory Guide	20
McFarland, Text-Book of Pathogenic Bacteria	15

Botany.
Bastin, Laboratory Exercises in Botany	20

Chemistry and Physics.
Brockway, Essentials of Physics	26
Wolff, Essentials of Chemistry	26

Children.
*An American Text-Book of Diseases of Children	8
Griffith, Care of the Baby	21
Powell, Essentials of Diseases of Children	26

Clinical Charts, Diet, and Diet Lists.
Hart, Diet in Sickness and in Health	22
Keen, Operation Blank	19
Laine, Temperature Chart	16
Meigs, Feeding in Early Infancy	14
Starr, Diets for Infants and Children	22
Thomas, Detachable Diet Lists, etc.	22

Diagnosis.
Cohen and Eshner, Essentials of Diagnosis	26
MacDonald, Surgical Diagnosis and Treatment	29
Vierordt and Stuart, Medical Diagnosis	10
Corwin, Essentials of the Physical Diagnosis of the Thorax	18

Dictionaries.
Keating and Hamilton, New Pronouncing Dictionary of Medicine	10
Morten, Nurses' Dictionary of Medical Terms	22
Saunders' Pocket Medical Lexicon	17

Ear.
Gleason, Essentials of Diseases of the Ear	26

Electricity.
Stewart and Lawrance, Essentials of Medical Electricity	26

Embryology.
Heisler, Text-Book of Embryology	29

Eye, Nose, and Throat.
De Schweinitz, Diseases of the Eye	14
Jackson and Gleason, Essentials of Diseases of Eye, Nose, and Throat	26
Kyle, Manual of Diseases of Nose and Throat	24

Genito-urinary.
Hyde, Syphilis and the Venereal Diseases	24
Martin, Essentials of Minor Surgery, Bandaging, and Venereal Diseases	26
Saundby, Renal and Urinary Diseases	27

Gynecology.
*An American Text-Book of Gynecology	9
Cragin, Essentials of Gynecology	26
Garrigues, Diseases of Women	18
Long, Syllabus of Gynecology	19

Histology.
Clarkson, Text-Book of Histology	15

Life Insurance.
Keating, How to Examine for Life Insurance	21

Materia Medica and Therapeutics.
*An American Text-Book of Applied Therapeutics	4
Butler, Text-Book of Materia Medica, Therapeutics, and Pharmacology	27
Cerna, Notes on the Newer Remedies	17
Griffin, Manual of Materia Medica and Therapeutics	24
Morris, Essentials of Materia Medica, etc.	26

	PAGE
Saunders' Pocket Medical Formulary	17
Stevens, Manual of Therapeutics	17
Thornton, Dose-Book and Prescription-Writing	24
Warren, Surgical Pathology and Therapeutics	11

Medical Jurisprudence.
Chapman, Medical Jurisprudence and Toxicology	24
Semple, Essentials of Legal Medicine, etc.	26

Medicine.
*An American Text-Book of Practice	7
*Gould and Pyle, Anomalies and Curiosities of Medicine	28
Lockwood, Manual of the Practice of Medicine	24
Morris, Essentials of the Practice of Medicine	26
Saunders' American Year-Book of Medicine and Surgery	30
Stevens, Manual of the Practice of Medicine	16

Nervous Diseases and Insanity.
Burr, Manual of Nervous Diseases	24
Shaw, Essentials of Nervous Diseases and Insanity	26

Nursing.
Griffith, Care of the Baby	21
Hampton, Nursing: Its Principles and Practice	21
Stoney, Practical Points in Private Nursing	13

Obstetrics.
*An American Text-Book of Obstetrics	5
Ashton, Essentials of Obstetrics	26
Boisliniere, Obstetric Accidents	20
Dorland, Manual of Obstetrics	24
Norris, Syllabus of Obstetrical Lectures	19

Pathology.
Semple, Essentials of Pathology and Morbid Anatomy	26
Senn, Pathology and Surgical Treatment of Tumors	11
Stengel, Manual of Pathology	24
Warren, Surgical Pathology and Therapeutics	11

Pharmacy.
Sayre, Essentials of Pharmacy	26

Physiology.
*An American Text-Book of Physiology	3
Hare, Essentials of Physiology	26
Raymond, Manual of Physiology	24
Stewart, A Manual of Physiology	15

Skiagraphy.
Rowland, Archives of Clinical Skiagraphy	16

Skin.
*Pictorial Atlas of Skin Diseases	12
Stelwagon, Essentials of Diseases of the Skin	26

Surgery.
*An American Text-Book of Surgery	6
Beck, Surgical Asepsis	24
DaCosta, Manual of Surgery	24
Keen, Operation Blank	19
MacDonald, Surgical Diagnosis and Treatment	29
Martin, Essentials of Surgery	26
Martin, Essentials of Minor Surgery, etc.	26
Pye, Elementary Bandaging and Surgical Dressing	27
Saunders' American Year-Book of Medicine and Surgery	30
Senn, Pathology and Surgical Treatment of Tumors	11
Senn, Syllabus of Surgery	19
Warren, Surgical Pathology and Therapeutics	11

Urine.
Wolff, Essentials of Examination of Urine	26

Miscellaneous.
Gross, Autobiography of	12
Saunders' New Aid Series of Manuals	23, 24
Saunders' Question Compends	25, 26
Thresh, Water and Water Supplies	15

For Sale by Subscription.

AN AMERICAN TEXT-BOOK OF PHYSIOLOGY. Edited by WILLIAM H. HOWELL, PH. D., M. D., Professor of Physiology in the Johns Hopkins University, Baltimore, Md. One handsome octavo volume of 1052 pages, fully illustrated. Prices: Cloth, $6.00 net; Sheep or Half-Morocco, $7.00 net.

This work is the most notable attempt yet made in America to combine in one volume the entire subject of Human Physiology by well-known teachers who have given especial study to that part of the subject upon which they write. The completed work represents the present status of the science of Physiology, particularly from the standpoint of the student of medicine and of the medical practitioner.

American teachers of physiology have not been altogether satisfied with the text-books at their disposal. The defects of most of the older books are that they have not kept pace with the rapid changes in modern physiology, while few if any of the newer books have been uniformly satisfactory in their treatment of all parts of this many-sided science. Indeed, the literature of experimental physiology is so great that it would seem to be almost impossible for any one teacher to keep thoroughly informed on all topics.

The collaboration of several teachers in the preparation of an elementary text-book of physiology is unusual, the almost invariable rule heretofore having been for a single author to write the entire book. One of the advantages to be derived from this collaboration method is that the more limited literature necessary for consultation by each author has enabled him to base his elementary account upon a comprehensive knowledge of the subject assigned to him; another, and perhaps the most important, advantage, is that the student gains the point of view of a number of teachers. In a measure he reaps the same benefit as would be obtained by following courses of instruction under different teachers. The different standpoints assumed, and the differences in emphasis laid upon the various lines of procedure, chemical, physical, and anatomical, should give the student a better insight into the methods of the science as it exists to-day. The work will also be found useful to many medical practitioners who may wish to keep in touch with the development of modern physiology.

The main divisions of the subject-matter are as follows: General Physiology of Muscle and Nerve—Secretion—Chemistry of Digestion and Nutrition—Movements of the Alimentary Canal, Bladder, and Ureter—Blood and Lymph—Circulation—Respiration—Animal Heat—Central Nervous System—Special Senses—Special Muscular Mechanisms—Reproduction—Chemistry of the Animal Body.

CONTRIBUTORS:

HENRY P. BOWDITCH, M. D.,
Professor of Physiology, Harvard Medical School.

JOHN G. CURTIS, M. D.,
Professor of Physiology, Columbia University, N. Y. (College of Physicians and Surgeons).

HENRY H. DONALDSON, Ph. D.,
Head-Professor of Neurology, University of Chicago.

W. H. HOWELL, Ph. D., M. D.,
Professor of Physiology, Johns Hopkins University.

FREDERIC S. LEE, Ph. D.,
Adjunct Prof. of Physiology, Columbia University, N. Y. (College of Physicians and Surgeons).

WARREN P. LOMBARD, M. D.,
Professor of Physiology, University of Michigan.

GRAHAM LUSK, Ph. D.,
Professor of Physiology, Yale Medical School.

W. T. PORTER, M. D.,
Assistant Professor of Physiology, Harvard Medical School.

EDWARD T. REICHERT, M. D.,
Professor of Physiology, University of Pennsylvania.

HENRY SEWALL, Ph. D., M. D.,
Professor of Physiology, Medical Department, University of Denver.

For Sale by Subscription.

AN AMERICAN TEXT-BOOK OF APPLIED THERAPEUTICS. For the Use of Practitioners and Students.
Edited by JAMES C. WILSON, M. D., Professor of the Practice of Medicine and of Clinical Medicine in the Jefferson Medical College. One handsome octavo volume of 1326 pages. Illustrated. Prices: Cloth, $7.00 net; Sheep or Half-Morocco, $8.00 net.

The arrangement of this volume has been based, so far as possible, upon modern pathologic doctrines, beginning with the intoxications and following with infections, diseases due to internal parasites, diseases of undetermined origin, and finally the disorders of the several bodily systems—digestive, respiratory, circulatory, renal, nervous, and cutaneous. It was thought proper to include also a consideration of the disorders of pregnancy.

The list of contributors comprises the names of many who have acquired distinction as practitioners and teachers of practice, of clinical medicine, and of the specialties.

CONTRIBUTORS:

Dr. I. E. Atkinson, Baltimore, Md.
Sanger Brown, Chicago, Ill.
John B. Chapin, Philadelphia, Pa.
William C. Dabney, Charlottesville, Va.
John Chalmers DaCosta, Phila., Pa.
I. N. Danforth, Chicago, Ill.
John L. Dawson, Jr., Charleston, S. C.
F. X. Dercum, Philadelphia, Pa.
George Dock, Ann Arbor, Mich.
Robert T. Edes, Jamaica Plain, Mass.
Augustus A. Eshner, Philadelphia, Pa.
J. T. Eskridge, Denver, Col.
F. Forchheimer, Cincinnati, O.
Carl Frese, Philadelphia, Pa.
Edwin E. Graham, Philadelphia, Pa.
John Guitéras, Philadelphia, Pa.
Frederick P. Henry, Philadelphia, Pa.
Guy Hinsdale, Philadelphia, Pa.
Orville Horwitz, Philadelphia, Pa.
W. W. Johnston, Washington, D. C.
Ernest Laplace, Philadelphia, Pa.
A. Laveran, Paris, France.

Dr. James Hendrie Lloyd, Phila., Pa.
John Noland Mackenzie, Balt., Md.
J. W. McLaughlin, Austin, Texas.
A. Lawrence Mason, Boston, Mass.
Charles K. Mills, Philadelphia, Pa.
John K. Mitchell, Philadelphia, Pa.
W. P. Northrup, New York City.
William Osler, Baltimore, Md.
Frederick A. Packard, Phila., Pa.
Theophilus Parvin, Philadelphia, Pa.
Beaven Rake, London, England.
E. O. Shakespeare, Philadelphia, Pa.
Wharton Sinkler, Philadelphia, Pa.
Louis Starr, Philadelphia, Pa.
Henry W. Stelwagon, Phila., Pa.
James Stewart, Montreal, Canada.
Charles G. Stockton, Buffalo, N. Y.
James Tyson, Philadelphia, Pa.
Victor C. Vaughan, Ann Arbor, Mich.
James T. Whittaker, Cincinnati, O.
J. C. Wilson, Philadelphia, Pa.

The articles, with two exceptions, are the contributions of American writers. Written from the standpoint of the practitioner, the aim of the work is to facilitate the application of knowledge to the prevention, the cure, and the alleviation of disease. The endeavor throughout has been to conform to the title of the book—Applied Therapeutics—to indicate the course of treatment to be pursued at the bedside, rather than to name a list of drugs that have been used at one time or another.

While the scientific superiority and the practical desirability of the metric system of weights and measures is admitted, it has not been deemed best to discard entirely the older system of figures, so that both sets have been given where occasion demanded.

For Sale by Subscription.

AN AMERICAN TEXT-BOOK OF OBSTETRICS. Edited by RICHARD C. NORRIS, M. D.; Art Editor, ROBERT L. DICKINSON, M. D. One handsome octavo volume of over 1000 pages, with nearly 900 colored and half-tone illustrations. Prices: Cloth, $7.00; Sheep or Half-Morocco, $8.00.

The advent of each successive volume of the *series* of the AMERICAN TEXT-BOOKS has been signalized by the most flattering comment from both the Press and the Profession. The high consideration received by these text-books, and their attainment to an authoritative position in current medical literature, have been matters of deep *international* interest, which finds its fullest expression in the demand for these publications from all parts of the civilized world.

In the preparation of the "AMERICAN TEXT-BOOK OF OBSTETRICS" the editor has called to his aid proficient collaborators whose professional prominence entitles them to recognition, and whose disquisitions exemplify **Practical Obstetrics**. While these writers were each assigned special themes for discussion, the correlation of the subject-matter is, nevertheless, such as ensures logical connection in treatment, the deductions of which thoroughly represent the latest advances in the science, and which elucidate *the best modern methods of procedure*.

The more conspicuous feature of the treatise is its wealth of illustrative matter. The production of the illustrations had been in progress for several years, under the personal supervision of Robert L. Dickinson, M. D., to whose artistic judgment and professional experience is due the **most sumptuously illustrated work of the period**. By means of the photographic art, combined with the skill of the artist and draughtsman, conventional illustration is superseded by rational methods of delineation.

Furthermore, the volume is a revelation as to the possibilities that may be reached in mechanical execution, through the unsparing hand of its publisher.

CONTRIBUTORS:

Dr. James C. Cameron.
Edward P. Davis.
Robert L. Dickinson.
Charles Warrington Earle.
James H. Etheridge.
Barton Cooke Hirst.
Henry J. Garrigues.
Charles Jewett.

Dr. Howard A. Kelly.
Richard C. Norris.
Chauncey D. Palmer.
Theophilus Parvin.
George A. Piersol.
Edward Reynolds.
Henry Schwarz.

"At first glance we are overwhelmed by the magnitude of this work in several respects, viz.: First, by the size of the volume, then by the array of eminent teachers in this department who have taken part in its production, then by the profuseness and character of the illustrations, and last, but not least, the conciseness and clearness with which the text is rendered. This is an entirely new composition, embodying the highest knowledge of the art as it stands to-day by authors who occupy the front rank in their specialty, and there are many of them. We cannot turn over these pages without being struck by the superb illustrations which adorn so many of them. We are confident that this most practical work will find instant appreciation by practitioners as well as students."—*New York Medical Times.*

Permit me to say that your American Text-Book of Obstetrics is the most magnificent medical work that I have ever seen. I congratulate you and thank you for this superb work, which alone is sufficient to place you first in the ranks of medical publishers.
With profound respect I am sincerely yours,
ALEX. J. C. SKENE.

For Sale by Subscription.

AN AMERICAN TEXT-BOOK OF SURGERY. Edited by WILLIAM W. KEEN, M. D., LL.D., and J. WILLIAM WHITE, M. D., PH. D. Forming one handsome royal-octavo volume of 1250 pages (10 × 7 inches), with 500 wood-cuts in text, and 37 colored and half-tone plates, many of them engraved from original photographs and drawings furnished by the authors. Prices: Cloth, $7.00; Sheep or Half-Morocco, $8.00 net.

SECOND EDITION, REVISED AND ENLARGED,
With a Section devoted to "The Use of the Röntgen Rays in Surgery."

The want of a text-book which could be used by the practitioner and at the same time be recommended to the medical student has been deeply felt, especially by teachers of surgery; hence, when it was suggested to a number of these that it would be well to unite in preparing a text-book of this description, great unanimity of opinion was found to exist, and the gentlemen below named gladly consented to join in its production.

Especial prominence has been given to Surgical Bacteriology, a feature which is believed to be unique in a surgical text-book in the English language. Asepsis and Antisepsis have received particular attention. The text is brought well up to date in such important branches as cerebral, spinal, intestinal, and pelvic surgery, the most important and newest operations in these departments being described and illustrated.

The text of the entire book has been submitted to all the authors for their mutual criticism and revision—an idea in book-making that is entirely new and original. The book as a whole, therefore, expresses on all the important surgical topics of the day the consensus of opinion of the eminent surgeons who have joined in its preparation. One of the most attractive features of the book is its illustrations. Very many of them are original and faithful reproductions of photographs taken directly from patients or from specimens, and the modern improvements in the art of engraving have enabled the publisher to produce illustrations which it is believed are superior to those in any similar work.

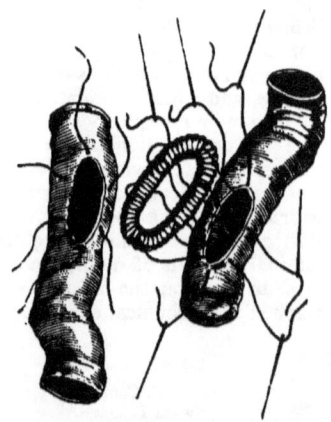

Specimen Illustration (largely reduced).

CONTRIBUTORS:

Dr. Charles H. Burnett, Philadelphia.
Phineas S. Conner, Cincinnati.
Frederic S. Dennis, New York.
William W. Keen, Philadelphia.
Charles B. Nancrede, Ann Arbor, Mich.
Roswell Park, Buffalo, N. Y.
Lewis S. Pilcher, Brooklyn, N. Y.

Dr. Nicholas Senn, Chicago.
Francis J. Shepherd, Montreal, Canada.
Lewis A. Stimson, New York.
William Thomson, Philadelphia.
J. Collins Warren, Boston.
J. William White, Philadelphia.

"If this text-book is a fair reflex of the present position of American surgery, we must admit it is of a very high order of merit, and that English surgeons will have to look very carefully to their laurels if they are to preserve a position in the van of surgical practice."—*London Lancet.*

"The soundness of the teachings contained in this work needs no stronger guarantee than is afforded by the names of its authors."—*Medical News,* Philadelphia.

For Sale by Subscription.

AN AMERICAN TEXT-BOOK ON THE THEORY AND PRACTICE OF MEDICINE. By American Teachers. Edited by WILLIAM PEPPER, M. D., LL.D., Provost and Professor of the Theory and Practice of Medicine and of Clinical Medicine in the University of Pennsylvania. Complete in two handsome royal-octavo volumes of about 1000 pages each, with illustrations to elucidate the text wherever necessary. Price per Volume: Cloth, $5.00 net; Sheep or Half-Morocco, $6.00 net.

VOLUME I. CONTAINS:

Hygiene.—Fevers (Ephemeral, Simple Continued, Typhus, Typhoid, Epidemic Cerebro-spinal Meningitis, and Relapsing).—Scarlatina, Measles, Rötheln, Variola, Varioloid, Vaccinia, Varicella, Mumps, Whooping-cough, Anthrax, Hydrophobia, Trichinosis, Actinomycosis, Glanders, and Tetanus.—Tuberculosis, Scrofula, Syphilis, Diphtheria, Erysipelas, Malaria, Cholera, and Yellow Fever.—Nervous, Muscular, and Mental Diseases.

VOLUME II. CONTAINS:

Urine (Chemistry and Microscopy).—Kidney and Lungs.—Air-passages (Larynx and Bronchi) and Pleura.—Pharynx, Œsophagus, Stomach and Intestines (including Intestinal Parasites), Heart, Aorta, Arteries and Veins.—Peritoneum, Liver, and Pancreas.—Diathetic Diseases (Rheumatism, Rheumatoid Arthritis, Gout, Lithæmia, and Diabetes).—Blood and Spleen.—Inflammation, Embolism, Thrombosis, Fever, and Bacteriology.

The articles are not written as though addressed to students in lectures, but are exhaustive descriptions of diseases, with the newest facts as regards Causation, Symptomatology, Diagnosis, Prognosis, and Treatment, including a large number of approved formulæ. The recent advances made in the study of the bacterial origin of various diseases are fully described, as well as the bearing of the knowledge so gained upon prevention and cure. The subjects of Bacteriology as a whole and of Immunity are fully considered in a separate section.

Methods of diagnosis are given the most minute and careful attention, thus enabling the reader to learn the very latest methods of investigation without consulting works specially devoted to the subject.

CONTRIBUTORS:

Dr. J. S. Billings, Philadelphia.
Francis Delafield, New York.
Reginald H. Fitz, Boston.
James W. Holland, Philadelphia.
Henry M. Lyman, Chicago.
William Osler, Baltimore.

Dr. William Pepper, Philadelphia.
W. Gilman Thompson, New York.
W. H. Welch, Baltimore.
James T. Whittaker, Cincinnati.
James C. Wilson, Philadelphia.
Horatio C. Wood, Philadelphia.

" We reviewed the first volume of this work, and said: 'It is undoubtedly one of the best text-books on the practice of medicine which we possess.' A consideration of the second and last volume leads us to modify that verdict and to say that the completed work is, in our opinion, the BEST of its kind it has ever been our fortune to see. It is complete, thorough, accurate, and clear. It is well written, well arranged, well printed, well illustrated, and well bound. It is a model of what the modern text-book should be."—*New York Medical Journal.*

" A library upon modern medical art. The work must promote the wider diffusion of sound knowledge."—*American Lancet.*

" A trusty counsellor for the practitioner or senior student, on which he may implicitly rely."—*Edinburgh Medical Journal.*

For Sale by Subscription.

AN AMERICAN TEXT-BOOK OF THE DISEASES OF CHILDREN.
By American Teachers. Edited by LOUIS STARR, M. D., assisted by THOMPSON S. WESTCOTT, M. D. In one handsome royal-8vo volume of 1190 pages, profusely illustrated with wood-cuts, half-tone and colored plates. Prices: Cloth, $7.00 net; Sheep or Half-Morocco, $8.00 net.

The plan of this work embraces a series of original articles written by some sixty well-known pædiatrists, representing collectively the teachings of the most prominent medical schools and colleges of America. The work is intended to be a PRACTICAL book, suitable for constant and handy reference by the practitioner and the advanced student.

One decided innovation is the large number of authors, nearly every article being contributed by a specialist in the line on which he writes. This, while entailing considerable labor upon the editors, has resulted in the publication of a work THOROUGHLY NEW AND ABREAST OF THE TIMES.

Especial attention has been given to the consideration of the latest accepted teaching upon the etiology, symptoms, pathology, diagnosis, and treatment of the disorders of children, with the introduction of many special formulæ and therapeutic procedures.

Special chapters embrace at unusual length the Diseases of the Eye, Ear, Nose and Throat, and the Skin; while the introductory chapters cover fully the important subjects of Diet, Hygiene, Exercise, Bathing, and the Chemistry of Food. Tracheotomy, Intubation, Circumcision, and such minor surgical procedures coming within the province of the medical practitioner, are carefully considered.

CONTRIBUTORS:

Dr. S. S. Adams, Washington.
John Ashhurst, Jr., Philadelphia.
A. D. Blackader, Montreal, Canada.
Dillon Brown, New York.
Edward M. Buckingham, Boston.
Charles W. Burr, Philadelphia.
W. E. Casselberry, Chicago.
Henry Dwight Chapin, New York.
W. S. Christopher, Chicago.
Archibald Church, Chicago.
Floyd M. Crandall, New York.
Andrew F. Currier, New York.
Roland G. Curtin, Philadelphia.
J. M. DaCosta, Philadelphia.
I. N. Danforth, Chicago.
Edward P. Davis, Philadelphia.
John B. Deaver, Philadelphia.
G. E. de Schweinitz, Philadelphia.
John Dorning, New York.
Charles Warrington Earle, Chicago.
Wm. A. Edwards, San Diego, Cal.
F. Forchheimer, Cincinnati.
J. Henry Fruitnight, New York.
Landon Carter Gray, New York.
J. P. Crozer Griffith, Philadelphia.
W. A. Hardaway, St. Louis.
M. P. Hatfield, Chicago.
Barton Cooke Hirst, Philadelphia.
H. Illoway, Cincinnati.
Henry Jackson, Boston.
Charles G. Jennings, Detroit.
Henry Koplik, New York.

Dr. Thomas S. Latimer, Baltimore.
Albert R. Leeds, Hoboken, N. J.
J. Hendrie Lloyd, Philadelphia.
George Roe Lockwood, New York.
Henry M. Lyman, Chicago.
Francis T. Miles, Baltimore.
Charles K. Mills, Philadelphia.
John H. Musser, Philadelphia.
Thomas R. Neilson, Philadelphia.
W. P. Northrup, New York.
William Osler, Baltimore.
Frederick A. Packard, Philadelphia.
William Pepper, Philadelphia.
Frederick Peterson, New York.
W. T. Plant, Syracuse, New York.
William M. Powell, Atlantic City.
B. Alexander Randall, Philadelphia.
Edward O. Shakespeare, Philadelphia.
F. C. Shattuck, Boston.
J. Lewis Smith, New York.
Louis Starr, Philadelphia.
M. Allen Starr, New York.
J. Madison Taylor, Philadelphia.
Charles W. Townsend, Boston.
James Tyson, Philadelphia.
W. S. Thayer, Baltimore.
Victor C. Vaughan, Ann Arbor, Mich.
Thompson S. Westcott, Philadelphia.
Henry R. Wharton, Philadelphia.
J. William White, Philadelphia.
J. C. Wilson, Philadelphia.

For Sale by Subscription.

AN AMERICAN TEXT-BOOK OF GYNECOLOGY, MEDICAL AND SURGICAL, for the use of Students and Practitioners. Edited by J. M. BALDY, M. D. Forming a handsome royal-octavo volume, with 360 illustrations in text and 37 colored and half-tone plates. Prices: Cloth, $6.00 net; Sheep or Half-Morocco, $7.00 net.

In this volume all anatomical descriptions, excepting those essential to a clear understanding of the text, have been omitted, the illustrations being largely depended upon to elucidate the anatomy of the parts. This work, which is thoroughly practical in its teachings, is intended, as its title implies, to be a working text-book for physicians and students. A clear line of treatment has been laid down in every case, and although no attempt has been made to discuss mooted points, still the most important of these have been noted and explained. The operations recommended are fully illustrated, so that the reader, having a picture of the procedure described in the text under his eye, cannot fail to grasp the idea. All extraneous matter and discussions have been carefully excluded, the attempt being made to allow no unnecessary details to cumber the text. The subject-matter is brought up to date at every point, and the work is as nearly as possible the combined opinions of the ten specialists who figure as the authors.

The work is well illustrated throughout with wood-cuts, half-tone and colored plates, mostly selected from the authors' private collections.

Specimen Illustration.

CONTRIBUTORS:

Dr. Henry T. Byford.
John M. Baldy.
Edwin Cragin.
J. H. Etheridge.
William Goodell.

Dr. Howard A. Kelly.
Florian Krug.
E. E. Montgomery.
William R. Pryor.
George M. Tuttle.

" The most notable contribution to gynecological literature since 1887, ... and the most complete exponent of gynecology which we have. No subject seems to have been neglected, ... and the gynecologist and surgeon and the general practitioner, who has any desire to practise diseases of women, will find it of practical value. In the matter of illustrations and plates the book surpasses anything we have seen."—*Boston Medical and Surgical Journal.*

A NEW PRONOUNCING DICTIONARY OF MEDICINE, with Phonetic Pronunciation, Accentuation, Etymology, etc. By JOHN M. KEATING, M. D., LL. D., Fellow of the College of Physicians of Philadelphia; Vice-President of the American Pædiatric Society; Ex-President of the Association of Life Insurance Medical Directors; Editor "Cyclopædia of the Diseases of Children," etc.; and HENRY HAMILTON, Author of a "A New Translation of Virgil's Æneid into English Rhyme;" Co-Author of "Saunders' Medical Lexicon," etc.; with the Collaboration of J. CHALMERS DACOSTA, M. D., and FREDERICK A. PACKARD, M. D. With an Appendix, containing Important Tables of Bacilli, Micrococci, Leucomaïnes, Ptomaïnes; Drugs and Materials used in Antiseptic Surgery; Poisons and their Antidotes; Weights and Measures; Thermometric Scales; New Official and Unofficial Drugs, etc. One volume of over 800 pages. Second Revised Edition. Prices: Cloth, $5.00; Sheep or Half-Morocco, $6.00 net; Half-Russia, $6.50 net, with Denison's Patent Ready-Reference Index; without Patent Index, Cloth, $4.00 net; Sheep or **Half-Morocco, $5.00 net.**

"I am much pleased with Keating's Dictionary, and shall take pleasure in recommending it to my classes."
HENRY M. LYMAN, M. D.,
Professor of Principles and Practice of Medicine, Rush Medical College, Chicago, Ill.

"I am convinced that it will be a very valuable adjunct to my study-table, convenient in size and sufficiently full for ordinary use." •
C. A. LINDSLEY, M. D.,
*Professor of Theory and Practice of Medicine, Medical Dept. Yale University;
Secretary Connecticut State Board of Health, New Haven, Conn.*

MEDICAL DIAGNOSIS. By Dr. OSWALD VIERORDT, Professor of Medicine at the University of Heidelberg. Translated, with additions, from the Second Enlarged German Edition, with the author's permission, by FRANCIS H. STUART, A. M., M. D. Third and Revised Edition. In one handsome royal-octavo volume of 700 pages, 178 fine wood-cuts in text, many of which are in colors. Prices: Cloth, $4.00 net; Sheep or Half-Morocco, $5.00 net; Half-Russia, $5.50 net.

In this work, as in no other hitherto published, are given full and accurate explanations of the phenomena observed at the bedside. It is distinctly a clinical work by a master teacher, characterized by thoroughness, fulness, and accuracy. It is a mine of information upon the points that are so often passed over without explanation. Especial attention has been given to the germ-theory as a factor in the origin of disease.

This valuable work is now published in German, English, Russian, and Italian. The issue of a third American edition within two years indicates the favor with which it has been received by the profession.

"Rarely is a book published with which a reviewer can find so little fault as with the volume before us. All the chapters are full, and leave little to be desired by the reader. Each particular item in the consideration of an organ or apparatus, which is necessary to determine a diagnosis of any disease of that organ, is mentioned; nothing seems forgotten. The chapters on diseases of the circulatory and digestive apparatus and nervous system are especially full and valuable. Notwithstanding a few minor errors in translating, which are of small importance to the accuracy of the rest of the volume, the reviewer would repeat that the book is one of the best—probably, *the best*—which has fallen into his hands. An excellent and comprehensive index of nearly one hundred pages closes the volume."—*University Medical Magazine*, Philadelphia.

PATHOLOGY AND SURGICAL TREATMENT OF TUMORS.
By N. Senn, M. D., Ph. D., LL. D., Professor of Surgery and of Clinical Surgery, Rush Medical College; Professor of Surgery, Chicago Polyclinic; Attending Surgeon to Presbyterian Hospital; Surgeon-in-Chief, St. Joseph's Hospital, Chicago. 710 pages, 515 engravings, including full-page colored plates. Prices: Cloth, $6.00 net; Half-Morocco, $7.00 net.

Books specially devoted to this subject are few, and in our text-books and systems of surgery this part of surgical pathology is usually condensed to a degree incompatible with its scientific and clinical importance. The author spent many years in collecting the material for this work, and has taken great pains to present it in a manner that should prove useful as a text-book for the student, a work of reference for the busy practitioner, and a reliable, safe guide for the surgeon. The more difficult operations are fully described and illustrated. More than *one hundred* of the illustrations are original, while the remainder were selected from books and medical journals not readily accessible to the student and the general practitioner.

"The appearance of such a work is most opportune. . . . In design and execution the work is such as will appeal to every student who appreciates the logical examination of facts and the practical exemplification of well-digested clinical observation."—*Medical Record*, New York.

"The most exhaustive of any recent book in English on this subject. It is well illustrated, and will doubtless remain as the principal monograph on the subject in our language for some years. The book is handsomely illustrated and printed, ... and the author has given a notable and lasting contribution to surgery."—*Journal of American Medical Association*, Chicago.

SURGICAL PATHOLOGY AND THERAPEUTICS.
By John Collins Warren, M. D., LL. D., Professor of Surgery, Medical Department Harvard University; Surgeon to the Massachusetts General Hospital, etc. A handsome octavo volume of 832 pages, with 136 relief and lithographic illustrations, 33 of which are printed in colors, and all of which were drawn by William J. Kaula from original specimens. Prices: Cloth, $6.00 net; Half-Morocco, $7.00 net.

"The volume is for the bedside, the amphitheatre, and the ward. It deals with things not as we see them through the microscope alone, but as the practitioner sees their effect in his patients; not only as they appear in and affect culture-media, but also as they influence the human body; and, following up the demonstrations of the nature of diseases, the author points out their logical treatment" (*New York Medical Journal*). "Indeed, the volume may be termed a modern medical classic, for such is the position to which it has already risen" (*Medical Age*, Detroit), "and is the handsomest specimen of bookmaking * * * that has ever been issued from the American medical press" (*American Journal of the Medical Sciences*, Philadelphia).

Without Exception, the Illustrations are the Best ever Seen in a Work of this Kind.

"A most striking and very excellent feature of this book is its illustrations. Without exception, from the point of accuracy and artistic merit, they are the best ever seen in a work of this kind. * * * Many of those representing microscopic pictures are so perfect in their coloring and detail as almost to give the beholder the impression that he is looking down the barrel of a microscope at a well-mounted section."—*Annals of Surgery*, Philadelphia.

AUTOBIOGRAPHY OF SAMUEL D. GROSS, M. D., Emeritus
Professor of Surgery in the Jefferson Medical College of Philadelphia, with Reminiscences of His Times and Contemporaries. Edited by his Sons, SAMUEL W. GROSS, M. D., LL.D., late Professor of Principles of Surgery and of Clinical Surgery in the Jefferson Medical College, and A. HALLER GROSS, A. M., of the Philadelphia Bar. Preceded by a Memoir of Dr. Gross, by the late Austin Flint, M. D., LL.D. In two handsome volumes, each containing over 400 pages, demy 8vo, extra cloth, gilt tops, with fine Frontispiece engraved on steel. Price, $5.00 net.

This autobiography, which was continued by the late eminent surgeon until within three months before his death, contains a full and accurate history of his early struggles, trials, and subsequent successes, told in a singularly interesting and charming manner, and embraces short and graphic pen-portraits of many of the most distinguished men—surgeons, physicians, divines, lawyers, statesmen, scientists, etc.—with whom he was brought in contact in America and in Europe; the whole forming a retrospect of more than three-quarters of a century.

"Dr. Gross . . . was perhaps the most eminent exponent of medical science that America has yet produced. His Autobiography, related as it is with a fulness and completeness seldom to be found in such works, is an interesting and valuable book. He comments on many things, especially, of course, on MEDICAL MEN AND MEDICAL PRACTICE, in a very interesting way. Details of professional life have also much in them that will be new."—*The Spectator*, London, England.

THE PICTORIAL ATLAS OF SKIN DISEASES AND SYPHILITIC AFFECTIONS (American Edition). Translation from the French. Edited by J. J. PRINGLE, M. B., F. R. C. P., Assistant Physician to, and Physician to the department for Diseases of the Skin at, the Middlesex Hospital, London. Photo-lithochromes from the famous models of dermatological and syphilitic cases in the Museum of the Saint-Louis Hospital, Paris, with explanatory wood-cuts and text. In 12 Parts, at $3.00 per Part. Parts 1 to 8 now ready.

"The plates are beautifully executed."—JONATHAN HUTCHINSON, M. D. (London Hospital).

"I strongly recommend this Atlas. The plates are exceedingly well executed, and will be of great value to all studying dermatology."—STEPHEN MACKENZIE, M. D. (London Hospital).

"The plates in this Atlas are remarkably accurate and artistic reproductions of *typical* examples of skin disease. The work will be of great value to the practitioner and student."—WILLIAM ANDERSON, M. D. (St. Thomas Hospital).

"If the succeeding parts of this Atlas are to be similar to Part 1, now before us, we have no hesitation in cordially recommending it to the favorable notice of our readers as one of the finest dermatological atlases with which we are acquainted."—*Glasgow Medical Journal*, Aug., 1895.

"Of all the atlases of skin diseases which have been published in recent years, the present one promises to be of greatest interest and value, especially from the standpoint of the general practitioner."—*American Medico-Surgical Bulletin*, Feb. 22, 1896.

"The introduction of explanatory wood-cuts in the text is a novel and most important feature which greatly furthers the easier understanding of the excellent plates, than which nothing, we venture to say, has been seen better in point of correctness, beauty, and general merit."—*New York Medical Journal*, Feb. 15, 1896.

"An interesting feature of the Atlas is the descriptive text, which is written for each picture by the physician who treated the case or at whose instigation the models have been made. We predict for this truly beautiful work a large circulation in all parts of the medical world where the names *St. Louis* and *Baretta* have preceded it."—*Medical Record*, N. Y., Feb. 1, 1896.

PRACTICAL POINTS IN NURSING. For Nurses in Private Practice. By EMILY A. M. STONEY, Graduate of the Training-School for Nurses, Lawrence, Mass.; Superintendent of the Training-School for Nurses, Carney Hospital, South Boston, Mass. 456 pages, handsomely illustrated with 73 engravings in the text, and 9 colored and half-tone plates. Cloth. Price, $1.75 net.

In this volume the author explains, in popular language and in the shortest possible form, the entire range of *private* nursing as distinguished from *hospital* nursing, and the nurse is instructed how best to meet the various emergencies of medical and surgical cases when distant from medical or surgical aid or when thrown on her own resources.

An especially valuable feature of the work will be found in the directions to the nurse how to *improvise* everything ordinarily needed in the sick-room, where the embarrassment of the nurse, owing to the want of proper appliances, is frequently extreme.

The work has been logically divided into the following sections:

I. The Nurse: her responsibilities, qualifications, equipment, etc.
II. The Sick-Room: its selection, preparation, and management.
III. The Patient: duties of the nurse in medical, surgical, obstetric, and gynecologic cases.
IV. Nursing in Accidents and Emergencies.
V. Nursing in Special Medical Cases.
VI. Nursing of the New-born and Sick Children.
VII. Physiology and Descriptive Anatomy.

The APPENDIX contains much information in compact form that will be found of great value to the nurse, including Rules for Feeding the Sick; Recipes for Invalid Foods and Beverages; Tables of Weights and Measures; Table for Computing the Date of Labor; List of Abbreviations; Dose-List; and a full and complete Glossary of Medical Terms and Nursing Treatment.

" There are few books intended for non-professional readers which can be so cordially endorsed by a medical journal as can this one."—*Therapeutic Gazette*, Aug. 15, 1896.

" This is a well-written, eminently practical volume, which covers the entire range of private nursing as distinguished from hospital nursing, and instructs the nurse how best to meet the various emergencies which may arise and how to prepare everything ordinarily needed in the illness of her patient."—*American Journal of Obstetrics and Diseases of Women and Children*, Aug., 1896.

" It is a work that the physician can place in the hands of his private nurses with the assurance of benefit."—*Ohio Medical Journal*, Aug., 1896.

A TEXT-BOOK OF BACTERIOLOGY, including the **Etiology and Prevention of Infective Diseases** and an account of **Yeasts and Moulds, Hæmatozoa, and Psorosperms.** By EDGAR M. CROOKSHANK, M. B., Professor of Comparative Pathology and Bacteriology, King's College, London. A handsome octavo volume of 700 pages, illustrated with 273 engravings in the text, and 22 original and colored plates. Price, $6.50 net.

This book, though nominally a Fourth Edition of Professor Crookshank's "MANUAL OF BACTERIOLOGY," is practically a new work, the old one having been reconstructed, greatly enlarged, revised throughout, and largely rewritten, forming a text-book for the Bacteriological Laboratory, for Medical Officers of Health, and for Veterinary Inspectors.

DISEASES OF THE EYE. A Hand-Book of Ophthalmic Practice.
By G. E. DE SCHWEINITZ, M. D., Professor of Ophthalmology in the Jefferson Medical College, Philadelphia, etc. A handsome royal-octavo volume of 679 pages, with 256 fine illustrations, many of which are original, and 2 chromo-lithographic plates. Prices: Cloth, $4.00 net; Sheep or Half-Morocco, $5.00 net.

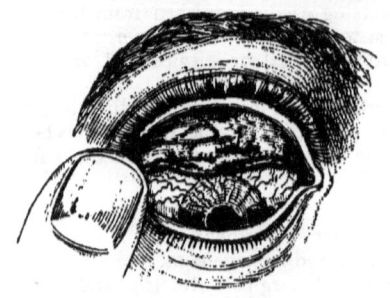

Specimen Illustration.

The object of this work is to present to the student, and to the practitioner who is beginning work in the fields of ophthalmology, a plain description of the optical defects and diseases of the eye. To this end special attention has been paid to the clinical side of the question; and the method of examination, the symptomatology leading to a diagnosis, and the treatment of the various ocular defects have been brought into prominence.

SECOND EDITION, REVISED AND GREATLY ENLARGED.

The entire book has been thoroughly revised. In addition to this general revision, special paragraphs on the following new matter have been introduced: Filamentous Keratitis, Blood-staining of the Cornea, Essential Phthisis Bulbi, Foreign Bodies in the Lens, Circinate Retinitis, Symmetrical Changes at the Macula Lutea in Infancy, Hyaline Bodies in the Papilla, Monocular Diplopia, Subconjunctival Injections of Germicides, Infiltration-Anæsthesia, and Sterilization of Collyria. Brief mention of Ophthalmia Nodosa, Electric Ophthalmia, and Angioid Streaks in the Retina also finds place. An Appendix has been added, containing a full description of the method of determining the corneal astigmatism with the ophthalmometer of Javal and Schiötz, and the rotations of the eyes with the tropometer of Stevens. The chapter on Operations has been enlarged and rewritten.

"A clearly written, comprehensive manual. . . . One which we can commend to students as a reliable text-book, written with an evident knowledge of the wants of those entering upon the study of this special branch of medical science."—*British Medical Journal.*

"The work is characterized by a lucidity of expression which leaves the reader in no doubt as to the meaning of the language employed. . . . We know of no work in which these diseases are dealt with more satisfactorily, and indications for treatment more clearly given, and in harmony with the practice of the most advanced ophthalmologists."—*Maritime Medical News.*

"It is hardly too much to say that for the student and practitioner beginning the study of Ophthalmology, it is the best single volume at present published."—*Medical News.*

"The latest and one of the best books on Ophthalmology. The book is thoroughly up to date, and is certainly a work which not only commends itself to the student, but is a ready reference for the busy practitioner."—*International Medical Magazine.*

FEEDING IN EARLY INFANCY. By ARTHUR V. MEIGS, M. D.
Bound in limp cloth, flush edges. Price, 25 cents net.

SYNOPSIS: Analyses of Milk—Importance of the Subject of Feeding in Early Infancy—Proportion of Casein and Sugar in Human Milk—Time to Begin Artificial Feeding of Infants—Amount of Food to be Administered at Each Feeding—Intervals between Feedings—Increase in Amount of Food at Different Periods of Infant Development—Unsuitableness of Condensed Milk as a Substitute for Mother's Milk—Objections to Sterilization or "Pasteurization" of Milk—Advances made in the Method of Artificial Feeding of Infants.

A TEXT-BOOK OF HISTOLOGY, DESCRIPTIVE AND PRACTICAL. For the Use of Students. By ARTHUR CLARKSON, M. B., C. M., Edin., formerly Demonstrator of Physiology in the Owen's College, Manchester; late Demonstrator of Physiology in the Yorkshire College, Leeds. Large 8vo, 554 pages, with 22 engravings in the text, and 174 beautifully colored original illustrations. Price, strongly bound in Cloth, $6.00 net.

The purpose of the writer in this work has been to furnish the student of Histology, in one volume, with both the descriptive and the practical part of the science. The first two chapters are devoted to the consideration of the general methods of Histology; subsequently, in each chapter, the structure of the tissue or organ is first systematically described, the student is then taken tutorially over the specimens illustrating it, and, finally, an appendix affords a short note of the methods of preparation.

TEXT-BOOK UPON THE PATHOGENIC BACTERIA. Specially written for Students of Medicine. By JOSEPH McFARLAND, M. D., Professor of Pathology and Bacteriology in the Medico-Chirurgical College of Philadelphia, etc. 359 pages, finely illustrated. Cloth. Price, $2.50 net.

The book presents a concise account of the technical procedures necessary in the study of Bacteriology. It describes the life-history of pathogenic bacteria, and the pathological lesions following invasions.

The work is intended to be a text-book for the medical student and for the practitioner who has had no recent laboratory training in this department of medical science. The instructions given as to needed apparatus, cultures, stainings, microscopic examinations, etc. are ample for the student's needs, and will afford to the physician much information that will interest and profit him.

"The author has succeeded admirably in presenting the essential details of bacteriological technics, together with a judiciously chosen summary of our present knowledge of pathogenic bacteria. . . . The work, we think, should have a wide circulation among English-speaking students of medicine."—*N. Y. Medical Journal*, April 4, 1896.

A MANUAL OF PHYSIOLOGY, with Practical Exercises. For Students and Practitioners. By G. N. STEWART, M. A., M. D., D. Sc., lately Examiner in Physiology, University of Aberdeen, and of the New Museums, Cambridge University; Professor of Physiology in the Western Reserve University, Cleveland, Ohio. Handsome octavo volume of 800 pages, with 278 illustrations in the text, and 5 colored plates. Price, Cloth, $3.50 net.

"It will make its way by sheer force of merit, and *amply deserves to do so. It is one of the very best English text-books on the subject.*"—*Lancet.*

"Of the many text-books of physiology published, we do not know of one that so nearly comes up to the ideal as does Prof. Stewart's volume."—*British Medical Journal.*

WATER AND WATER SUPPLIES. By JOHN C. THRESH, D. Sc., M. B., D. P. H., Lecturer on Public Health, King's College, London; Editor of the "Journal of State Medicine," etc. 12mo, 438 pages, illustrated. Handsomely bound in Cloth, with gold side and back stamps. Price, $2.25 net.

ARCHIVES OF CLINICAL SKIAGRAPHY. By SYDNEY ROWLAND, B. A., Camb., late Scholar of Downing College, Cambridge, and Shuter Scholar of St. Bartholomew's Hospital, London; Special Commissioner to "British Medical Journal" for the Investigation of the Applications of the New Photography to Medicine and Surgery. A series of collotype illustrations, with descriptive text, illustrating the applications of the New Photography to Medicine and Surgery. Price, per Part, $1.00. Parts I. to III. now ready.

The object of this publication is to put on record in permanent form some of the most striking applications of the new photography to the needs of Medicine and Surgery.

The progress of this new art has been so rapid that, although Prof. Röntgen's discovery is only a thing of yesterday, it has already taken its place among the approved and accepted aids to diagnosis.

ESSENTIALS OF ANATOMY AND MANUAL OF PRACTICAL DISSECTION, containing "Hints on Dissection." By CHARLES B. NANCREDE, M. D., Professor of Surgery and Clinical Surgery in the University of Michigan, Ann Arbor; Corresponding Member of the Royal Academy of Medicine, Rome, Italy; late Surgeon Jefferson Medical College, etc. Fourth and revised edition. Post 8vo, over 500 pages, with handsome full-page lithographic plates in colors, and over 200 illustrations. Price: Extra Cloth (or Oilcloth for the dissection-room), $2.00 net.

No pains nor expense has been spared to make this work the most exhaustive yet concise Student's Manual of Anatomy and Dissection ever published, either in America or in Europe. The colored plates are designed to aid the student in dissecting the muscles, arteries, veins, and nerves. The wood-cuts have all been specially drawn and engraved, and an Appendix added containing 60 illustrations representing the structure of the entire human skeleton, the whole being based on the eleventh edition of Gray's *Anatomy*.

A MANUAL OF PRACTICE OF MEDICINE. By A. A. STEVENS, A. M., M. D., Instructor of Physical Diagnosis in the University of Pennsylvania, and Demonstrator of Pathology in the Woman's Medical College of Philadelphia. Specially intended for students preparing for graduation and hospital examinations. Post 8vo, 512 pages. Illustrated. Price, $2.50.

FOURTH EDITION, REVISED AND ENLARGED.

Contributions to the science of medicine have poured in so rapidly during the last quarter of a century that it is well-nigh impossible for the student, with the limited time at his disposal, to master elaborate treatises or to cull from them that knowledge which is absolutely essential. From an extended experience in teaching, the author has been enabled, by classification, to group allied symptoms, and by the elimination of theories and redundant explanations to bring within a comparatively small compass a complete outline of the practice of medicine.

TEMPERATURE CHART. Prepared by D. T. LAINÉ, M. D. Size 8 x 13½ inches. Price, per pad of 25 charts, 50 cents net.

A conveniently arranged chart for recording Temperature, with columns for daily amounts of Urinary and Fecal Excretions, Food, Remarks, etc. On the back of each chart is given in full the method of Brand in the treatment of Typhoid Fever.

MANUAL OF MATERIA MEDICA AND THERAPEUTICS. By
A. A. STEVENS, A. M., M. D., Instructor of Physical Diagnosis in the University of Pennsylvania, and Demonstrator of Pathology in the Woman's Medical College of Philadelphia. 445 pages. Price, Cloth, $2.25.

SECOND EDITION, REVISED.

This wholly new volume, which is based on the last edition of the *Pharmacopœia*, comprehends the following sections: Physiological Action of Drugs; Drugs; Remedial Measures other than Drugs; Applied Therapeutics; Incompatibility in Prescriptions; Table of Doses; Index of Drugs; and Index of Diseases; the treatment being elucidated by more than two hundred formulæ.

NOTES ON THE NEWER REMEDIES: their Therapeutic Applications and Modes of Administration. By DAVID CERNA, M.D., PH.D.,
Demonstrator of and Lecturer on Experimental Therapeutics in the University of Pennsylvania. Post 8vo, 253 pages. Price, $1.25.

SECOND EDITION, RE-WRITTEN AND GREATLY ENLARGED.

The work takes up in alphabetical-order all the newer remedies, giving their physical properties, solubility, therapeutic applications, administration, and chemical formula.

SAUNDERS' POCKET MEDICAL FORMULARY. BY WILLIAM
M. POWELL, M. D., Attending Physician to the Mercer House for Invalid Women at Atlantic City. Containing 1750 Formulæ, selected from several hundred of the best-known authorities. Forming a handsome and convenient pocket companion of nearly 300 printed pages, with blank leaves for additions; with an Appendix containing Posological Table, Formulæ and Doses for Hypodermic Medication, Poisons and their Antidotes, Diameters of the Female Pelvis and Fœtal Head, Obstetrical Table, Diet List for Various Diseases, Materials and Drugs used in Antiseptic Surgery, Treatment of Asphyxia from Drowning, Surgical Remembrancer, Tables of Incompatibles, Eruptive Fevers, Weights and Measures, etc. Third edition, revised and greatly enlarged. Handsomely bound in morocco, with side index, wallet, and flap. Price, $1.75 net.

"This little book, that can be conveniently carried in the pocket, contains an immense amount of material. It is very useful, and as the name of the author of each prescription is given is unusually reliable."—*New York Medical Record.*

SAUNDERS' POCKET MEDICAL LEXICON; or, Dictionary of Terms and Words used in Medicine and Surgery. By JOHN M.
KEATING, M. D., Editor of "Cyclopædia of Diseases of Children," etc.; Author of the "New Pronouncing Dictionary of Medicine," and HENRY HAMILTON, Author of "A New Translation of Virgil's Æneid into English Verse;" Co-Author of a "New Pronouncing Dictionary of Medicine." A new and revised edition. 32mo, 282 pages. Prices: Cloth, 75 cents; Leather Tucks, $1.00.

"Remarkably accurate in terminology, accentuation, and definition."—*Journal of American Medical Association.*

DISEASES OF WOMEN.

By HENRY J. GARRIGUES, A. M., M. D., Professor of Obstetrics in the New York Post-Graduate Medical School and Hospital; Gynæcologist to St. Mark's Hospital, and to the German Dispensary, etc., New York City. One octavo volume of nearly 700 pages, illustrated by 300 wood-cuts and colored plates. Prices: Cloth, $4.00 net; Sheep, $5.00 net.

A PRACTICAL work on gynæcology for the use of students and practitioners, written in a terse and concise manner. The importance of a thorough knowledge of the anatomy of the female pelvic organs has been fully recognized by the author, and considerable space has been devoted to the subject. The chapters on Operations and on Treatment are thoroughly modern, and are based upon the large hospital and private practice of the author. The text is elucidated by a large number of illustrations and colored plates, many of them being original, and forming a complete atlas for studying *embryology* and the *anatomy* of the *female genitalia*, besides exemplifying, whenever needed, morbid conditions, instruments, apparatus, and operations.

EXCERPT OF CONTENTS.

Development of the Female Genitals.—Anatomy of the Female Pelvic Organs.—Physiology.—Puberty.—Menstruation and Ovulation.—Copulation.—Fecundation.—The Climacteric.—Etiology in General.—Examinations in General.—Treatment in General.—Abnormal Menstruation and Metrorrhagia.—Leucorrhea.—Diseases of the Vulva.—Diseases of the Perineum.—Diseases of the Vagina.—Diseases of the Uterus.—Diseases of the Fallopian Tubes.—Diseases of the Ovaries.—Diseases of the Pelvis.—Sterility.

The reception accorded to this work has been most flattering. In the short period which has elapsed since its issue, it has been adopted and recommended as a text-book by more than 60 of the Medical Schools and Universities of the United States and Canada.

"One of the best text-books for students and practitioners which has been published in the English language; it is condensed, clear, and comprehensive. The profound learning and great clinical experience of the distinguished author find expression in this book in a most attractive and instructive form. Young practitioners, to whom experienced consultants may not be available, will find in this book invaluable counsel and help."

THAD. A. REAMY, M. D., LL.D.,
Professor of Clinical Gynecology, Medical College of Ohio; Gynecologist to the Good Samaritan and to the Cincinnati Hospitals.

ESSENTIALS OF PHYSICAL DIAGNOSIS OF THE THORAX.

By ARTHUR M. CORWIN, A. M., M. D., Demonstrator of Physical Diagnosis in the Rush Medical College, Chicago; Attending Physician to the Central Free Dispensary, Department of Rhinology, Laryngology, and Diseases of the Chest. 200 pages. Illustrated. Cloth, flexible covers. Price, $1.25 net.

This book was originally published for the use of students, but its rapid absorption by the practitioner made it appear that a wider field had been reached. In this edition the author has added to his revision of the text a section setting forth the signs found in each of the diseases of the chest, thereby increasing its value to the general practitioner for post-graduate study.

"It is excellent. The student who shall use it as his guide to the careful study of physical exploration upon normal and abnormal subjects can scarcely fail to acquire a good working knowledge of the subject."—*Philadelphia Polyclinic.*

SYLLABUS OF OBSTETRICAL LECTURES in the Medical Department, University of Pennsylvania. By RICHARD C. NORRIS, A. M., M. D., Demonstrator of Obstetrics in the University of Pennsylvania. Third edition, thoroughly revised and enlarged. Crown 8vo. Price, Cloth, interleaved for notes, $2.00 net.

"This work is so far superior to others on the same subject that we take pleasure in calling attention briefly to its excellent features. It covers the subject thoroughly, and will prove invaluable both to the student and the practitioner. The author has introduced a number of valuable hints which would only occur to one who was himself an experienced teacher of obstetrics. The subject-matter is clear, forcible, and modern. We are especially pleased with the portion devoted to the practical duties of the accoucheur, care of the child, etc. The paragraphs on antiseptics are admirable; there is no doubtful tone in the directions given. No details are regarded as unimportant; no minor matters omitted. We venture to say that even the old practitioner will find useful hints in this direction which he cannot afford to despise."—*Medical Record.*

A SYLLABUS OF GYNECOLOGY, arranged in conformity with "An American Text-Book of Gynecology." By J. W. LONG, M. D., Professor of Diseases of Women and Children, Medical College of Virginia, etc. Price, Cloth (interleaved), $1.00 net.

Based upon the teaching and methods laid down in the larger work, this will not only be useful as a supplementary volume, but to those who do not already possess the Text-Book it will also have an independent value as an aid to the practitioner in gynecological work, and to the student as a guide in the lecture-room, as the subject is presented in a manner systematic, succinct, and practical.

A SYLLABUS OF LECTURES ON THE PRACTICE OF SURGERY, arranged in conformity with "An American Text-Book of Surgery." By NICHOLAS SENN, M. D., PH. D., Professor of Surgery in Rush Medical College, Chicago, and in the Chicago Polyclinic. Price, $2.00.

This excellent work of its eminent author, himself one of the contributors to "An American Text-Book of Surgery," will prove of exceptional value to the advanced student who has adopted that work as his text-book. It is not only the syllabus of an unrivalled course of surgical practice, but it is also an epitome of, or supplement to the larger work.

AN OPERATION BLANK, with Lists of Instruments, etc. required in Various Operations. Prepared by W. W. KEEN, M. D., LL.D., Professor of Principles of Surgery in the Jefferson Medical College, Philadelphia. Price per pad, containing Blanks for fifty operations, 50 cents net.

SECOND EDITION, REVISED FORM.

A convenient blank (suitable for all operations), giving complete instructions regarding necessary preparation of patient, etc., with a full list of dressings and medicines to be employed. On the back of each blank is a list of instruments used—viz. general instruments, etc., required for all operations; and special instruments for surgery of the brain and spine, mouth and throat, abdomen, rectum, male and female genito-urinary organs, the bones, etc. The whole forming a neat pad, arranged for hanging on the wall of a surgeon's office or in the hospital operating-room.

LABORATORY EXERCISES IN BOTANY.

By EDSON S. BASTIN, M. A., Professor of Materia Medica and Botany in the Philadelphia College of Pharmacy. Octavo volume of 536 pages, with 87 plates. Price, Cloth, $2.50.

This work is intended for the beginner and the advanced student, and it fully covers the structure of flowering plants, roots, ordinary stems, rhizomes, tubers, bulbs, leaves, flowers, fruits, and seeds. Particular attention is given to the gross and microscopical structure of plants, and to those used in medicine. The illustrations fully elucidate the text, and the complete index facilitates reference.

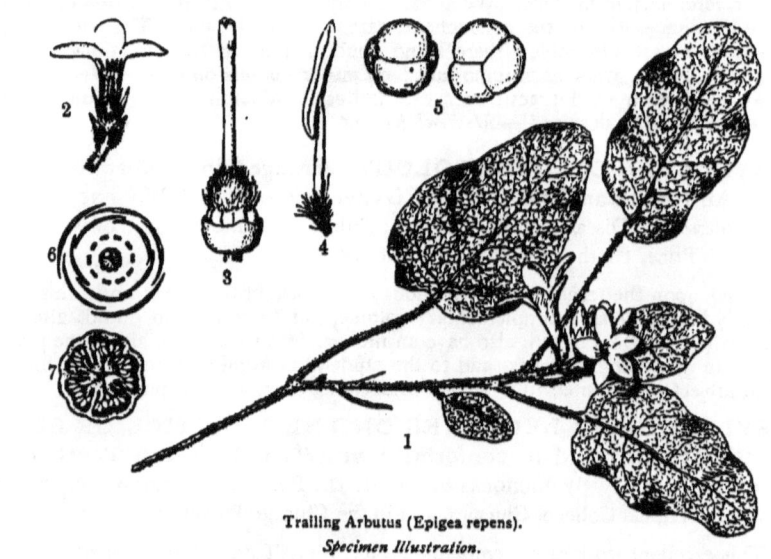

Trailing Arbutus (Epigea repens).
Specimen Illustration.

LABORATORY GUIDE FOR THE BACTERIOLOGIST.

By LANGDON FROTHINGHAM, M. D. V., Assistant in Bacteriology and Veterinary Science, Sheffield Scientific School, Yale University. Illustrated. Price, Cloth, 75 cents.

The technical methods involved in bacteria-culture, methods of staining, and microscopical study are fully described and arranged as simply and concisely as possible. The book is especially intended for use in laboratory work.

OBSTETRIC ACCIDENTS, EMERGENCIES, AND OPERATIONS.

By L. CH. BOISLINIERE, M. D., late Emeritus Professor of Obstetrics in the St. Louis Medical College. 381 pages, handsomely illustrated. Price, $2.00 net.

"For the use of the practitioner who, when away from home, has not the opportunity of consulting a library or of calling a friend in consultation. He then, being thrown upon his own resources, will find this book of benefit in guiding and assisting him in emergencies."

HOW TO EXAMINE FOR LIFE INSURANCE. By JOHN M.
KEATING, M. D., Fellow of the College of Physicians and Surgeons of Philadelphia; Vice-President of the American Pædiatric Society; Ex-President of the Association of Life Insurance Medical Directors. Royal 8vo, 211 pages, with two large half-tone illustrations, and a plate prepared by Dr. McClellan from special dissections; also, numerous cuts to elucidate the text. Price, in Cloth, $2.00 net.

"This is by far the most useful book which has yet appeared on insurance examination, a subject of growing interest and importance. Not the least valuable portion of the volume is Part II., which consists of instructions issued to their examining physicians by twenty-four representative companies of this country. As the proofs of these instructions were corrected by the directors of the companies, they form the latest instructions obtainable. If for these alone the book should be at the right hand of every physician interested in this special branch of medical science."—*The Medical News*, Philadelphia.

THE CARE OF THE BABY. By J. P. CROZER GRIFFITH, M. D., Clinical Professor of Diseases of Children, University of Pennsylvania; Physician to the Children's Hospital, Philadelphia, etc. 392 pages, with 67 illustrations in the text, and 5 plates. 12mo. Price, $1.50.

A reliable guide not only for mothers, but also for medical students and practitioners whose opportunities for observing children have been limited.

"The whole book is characterized by rare good sense, and is evidently written by a master hand. It can be read with benefit not only by mothers, but by medical students and by any practitioners who have not had large opportunities for observing children."—*American Journal of Obstetrics*, July, 1895.

"The best book for the use of the young mother with which we are acquainted. . . . There are very few general practitioners who could not read the book through with advantage."—*Archives of Pediatrics*, Aug., 1895.

"No better book of its kind has come under our notice for some time. Although intended primarily for mothers and nurses, it will well repay perusal by medical students."—*Birmingham Medical Review*, Oct., 1895.

"This is one of the best works of its kind that has been presented to the people for many a day."—*Maryland Medical Journal*, Aug. 13, 1895.

NURSING: ITS PRINCIPLES AND PRACTICE. By ISABEL ADAMS
HAMPTON, Graduate of the New York Training School for Nurses attached to Bellevue Hospital; Superintendent of Nurses, and Principal of the Training School for Nurses, Johns Hopkins Hospital, Baltimore, Md.; late Superintendent of Nurses, Illinois Training School for Nurses, Chicago, Ill. In one very handsome 12mo volume of 484 pages, profusely illustrated. Price, Cloth, $2.00 net.

This original work on the important subject of nursing is at once comprehensive and systematic. It is written in a clear, accurate, and readable style, suitable alike to the student and the lay reader. Such a work has long been a desideratum with those intrusted with the management of hospitals and the instruction of nurses in training-schools. It is also of especial value to the graduate nurse who desires to acquire a practical working knowledge of the care of the sick and the hygiene of the sick-room.

NURSE'S DICTIONARY of Medical Terms and Nursing Treatment, containing Definitions of the Principal Medical and Nursing Terms and Abbreviations; of the Instruments, Drugs, Diseases, Accidents, Treatments, Physiological Names, Operations, Foods, Appliances, etc. encountered in the ward or in the sick-room. Compiled for the use of nurses. By HONNOR MORTEN, Author of "How to Become a Nurse," "Sketches of Hospital Life," etc. 16mo, 140 pages. Price, Cloth, $1.00.

This little volume is intended merely as a small reference-book which can be consulted at the bedside or in the ward. It gives sufficient explanation to the nurse to enable her to comprehend a case until she has leisure to look up larger and fuller works on the subject.

DIET IN SICKNESS AND IN HEALTH. By MRS. ERNEST HART, formerly Student of the Faculty of Medicine of Paris and of the London School of Medicine for Women; with an INTRODUCTION by Sir Henry Thompson, F. R. C. S., M. D., London. 220 pages; illustrated. Price, Cloth, $1.50.

Useful to those who have to nurse, feed, and prescribe for the sick. . . . In each case the accepted causation of the disease and the reasons for the special diet prescribed are briefly described. Medical men will find the dietaries and recipes practically useful, and likely to save them trouble in directing the dietetic treatment of patients.

"We recommend it cordially to the attention of all practitioners; . . . . both to them and to their patients it may be of the greatest service."—*Medical Journal*, New York.

DIETS FOR INFANTS AND CHILDREN IN HEALTH AND IN DISEASE. By LOUIS STARR, M. D., Editor of "An American Text-Book of the Diseases of Children." 230 blanks (pocket-book size), perforated and neatly bound in flexible morocco. Price, $1.25 net.

The first series of blanks are prepared for the first seven months of infant life; each blank indicates the ingredients, but not the *quantities*, of the food, the latter directions being left for the physician. After the seventh month, modifications being less necessary, the diet lists are printed in full. *Formulæ* for the preparation of diluents and foods are appended.

DIET LISTS AND SICK-ROOM DIETARY. By JEROME B. THOMAS, M. D., Visiting Physician to the Home for Friendless Women and Children and to the Newsboys' Home; Assistant Visiting Physician to the Kings County Hospital; Assistant Bacteriologist, Brooklyn Health Department. Price, $1.50. Send for sample sheet.

There is here offered, in portable form, as an efficient aid to the better practice of Therapeutics, a collection of detachable Diet Lists and a Sick-room Dietary. It meets a want, for the busy practitioner has but little time to write out *Systems of Diet* appropriate to his patients, or to describe the preparation of their food. Compiled from the most modern works on dietetics, the Dietary offers a variety of easily-digested foods.

"A convenience that will be appreciated by the physician."—*Medical Journal*, New York.

"The work is an excellent one, and ought to be welcomed by physician, patient, and nurse alike."—*Indian Lancet*, Calcutta.

Practical, Exhaustive, Authoritative.

SAUNDERS'
NEW AID SERIES OF MANUALS.
FOR
STUDENTS AND PRACTITIONERS.

MR. SAUNDERS is pleased to announce the successful issue of several volumes of his **NEW AID SERIES OF MANUALS**, which have received the most flattering commendations from Students and Practitioners and the Press. As publisher of the STANDARD SERIES OF QUESTION COMPENDS, and through intimate relations with leading members of the medical profession, Mr. Saunders has been enabled to study progressively the essential *desiderata* in practical "self-helps" for students and physicians.

This study has manifested that, while the published "Question Compends" earn the highest appreciation of students, whom they serve in reviewing their studies preparatory to examination, there is special need of thoroughly reliable handbooks on the leading branches of Medicine and Surgery, each subject being compactly and authoritatively written, and exhaustive in detail, without the introduction of *cases* and foreign subject-matter which so largely expand ordinary text-books.

The **Saunders Aid Series will not merely be condensations from present literature, but will be ably written by well-known authors and practitioners, most of them being teachers in representative American Colleges.** This *new series*, therefore, will form an admirable collection of advanced lectures, which will be invaluable aids to students in reading and in comprehending the contents of "recommended" works.

Each Manual will further be distinguished by the beauty of the *new* type; by the quality of the paper and printing; by the copious use of illustrations; by the attractive binding in cloth; **and by the extremely low price at which they will be sold.**

Saunders' New Aid Series of Manuals.

VOLUMES PUBLISHED.

PHYSIOLOGY, by JOSEPH HOWARD RAYMOND, A. M., M. D., Professor of Physiology and Hygiene and Lecturer on Gynecology in the Long Island College Hospital; Director of Physiology in the Hoagland Laboratory; formerly Lecturer on Physiology and Hygiene in the Brooklyn Normal School for Physical Education; Ex-Vice-President of the American Public Health Association; Ex-Health Commissioner, City of Brooklyn, etc. Illustrated. $1.25 net.

SURGERY, General and Operative, by JOHN CHALMERS DACOSTA, M. D., Demonstrator of Surgery, Jefferson Medical College, Philadelphia; Chief Assistant Surgeon, Jefferson Medical College Hospital; Surgical Registrar, Philadelphia Hospital, etc. 188 illustrations and 13 plates. (Double number.) $2.50 net.

DOSE-BOOK AND MANUAL OF PRESCRIPTION-WRITING, by E. Q. THORNTON, M. D., Demonstrator of Therapeutics, Jefferson Medical College, Philadelphia. Illustrated. Price, cloth, $1.25 net.

SURGICAL ASEPSIS, by CARL BECK, M. D., Surgeon to St. Mark's Hospital and to the New York German Poliklinik, etc. Illustrated. Price, cloth, $1.25 net.

MEDICAL JURISPRUDENCE, by HENRY C. CHAPMAN, M. D., Professor of Institutes of Medicine and Medical Jurisprudence in the Jefferson Medical College of Philadelphia; Member of the College of Physicians of Philadelphia, of the Academy of Natural Sciences, of the American Philosophical Society, and of the Zoological Society of Philadelphia. Illustrated. $1.50 net.

SYPHILIS AND THE VENEREAL DISEASES, by JAMES NEVINS HYDE, M. D., Professor of Skin and Venereal Diseases, and FRANK H. MONTGOMERY, M. D., Lecturer on Dermatology and Genito-Urinary Diseases, in Rush Medical College, Chicago. Profusely Illustrated. (Double number.) $2.50 net.

PRACTICE OF MEDICINE, by GEORGE ROE LOCKWOOD, M. D., Professor of Practice in the Woman's Medical College of the New York Infirmary; Instructor of Physical Diagnosis of the Medical Department of Columbia College; Attending Physician to the Colored Hospital; Pathologist to the French Hospital; Member of the New York Academy of Medicine, of the Pathological Society, of the Clinical Society, etc. Illustrated. (Double number.) $2.50 net.

MANUAL OF ANATOMY, by IRVING S. HAYNES, M. D., Adjunct Professor of Anatomy and Demonstrator of Anatomy, Medical Department of the New York University, etc. Beautifully Illustrated. (Double number.) Price, $2.50 net.

MANUAL OF OBSTETRICS, by W. A. NEWMAN DORLAND, M. D., Asst. Demonstrator of Obstetrics, University of Pennsylvania; Chief of Gynecological Dispensary, Pennsylvania Hospital; Member of Philadelphia Obstetrical Society, etc. Profusely illustrated. (Double number.) Price, $2.50 net.

DISEASES OF WOMEN, by J. BLAND SUTTON, F. R. C. S., Asst. Surgeon to Middlesex Hospital, and Surgeon to Chelsea Hospital, London; and ARTHUR E. GILES, M. D., B. Sc. Lond., F. R. C. S. Edin., Asst. Surgeon to Chelsea Hospital, London. 436 pages, handsomely illustrated. (Double number.) Price, $2.50 net.

VOLUMES IN PREPARATION.

NOSE AND THROAT, by D. BRADEN KYLE, M. D., Chief Laryngologist of the St. Agnes Hospital, Philadelphia; Bacteriologist of the Orthopædic Hospital and Infirmary for Nervous Diseases; Instructor in Clinical Microscopy and Assistant Demonstrator of Pathology in the Jefferson Medical College, etc.

NERVOUS DISEASES, by CHARLES W. BURR, M. D., Clinical Professor of Nervous Diseases, Medico-Chirurgical College, Philadelphia; Pathologist to the Orthopædic Hospital and Infirmary for Nervous Diseases; Visiting Physician to the St. Joseph Hospital, etc.

*** There will be published in the same series, at close intervals, carefully-prepared works on various subjects, by prominent specialists.

SAUNDERS' QUESTION COMPENDS.
Arranged in Question and Answer Form.

THE LATEST, CHEAPEST, AND BEST ILLUSTRATED SERIES OF COMPENDS EVER ISSUED.

Now the Standard Authorities in Medical Literature
WITH
Students and Practitioners in every City of the United States and Canada.

THE REASON WHY

They are the advance guard of "Student's Helps"—that DO HELP; they are the leaders in their special line, *well and authoritatively written by able men, who, as teachers in the large colleges, know exactly what is wanted by a student preparing for his examinations.* The judgment exercised in the selection of authors is fully demonstrated by their professional elevation. Chosen from the ranks of Demonstrators, Quiz-masters, and Assistants, most of them have become Professors and Lecturers in their respective colleges.

Each book is of convenient size (5 × 7 inches), containing on an average 250 pages, profusely illustrated, and elegantly printed in clear, readable type, on fine paper.

The entire series, numbering twenty-three volumes, has been kept thoroughly revised and enlarged when necessary, many of them being in their fourth and fifth editions.

TO SUM UP.

Although there are numerous other Quizzes, Manuals, Aids, etc. in the market, none of them approach the "Blue Series of Question Compends;" and the claim is made for the following points of excellence :

1. Professional distinction and reputation of authors.
2. Conciseness, clearness, and soundness of treatment.
3. Size of type and quality of paper and binding.

⁎ **Any of these Compends will be mailed on receipt of price (see over for List).**

Saunders' Question-Compend Series.

☞ **Price, Cloth, $1.00 per copy, except when otherwise noted.**

1. **ESSENTIALS OF PHYSIOLOGY.** 3d edition. Illustrated. Revised and enlarged. By H. A. HARE, M. D. (Price, $1.00 net.)
2. **ESSENTIALS OF SURGERY.** 5th edition, with an Appendix on Antiseptic Surgery. 90 illustrations. By EDWARD MARTIN, M. D.
3. **ESSENTIALS OF ANATOMY.** 5th edition, with an Appendix. 180 illustrations. By CHARLES B. NANCREDE, M. D.
4. **ESSENTIALS OF MEDICAL CHEMISTRY, ORGANIC AND INORGANIC.** 4th edition, revised, with an Appendix. By LAWRENCE WOLFF, M. D.
5. **ESSENTIALS OF OBSTETRICS.** 3d edition, revised and enlarged. 75 illustrations. By W. EASTERLY ASHTON, M. D.
6. **ESSENTIALS OF PATHOLOGY AND MORBID ANATOMY.** 6th thousand. 46 illustrations. By C. E. ARMAND SEMPLE, M. D.
7. **ESSENTIALS OF MATERIA MEDICA, THERAPEUTICS, AND PRESCRIPTION-WRITING.** 4th edition. By HENRY MORRIS, M. D.
8, 9. **ESSENTIALS OF PRACTICE OF MEDICINE.** By HENRY MORRIS, M. D. An Appendix on URINE EXAMINATION. Illustrated. By LAWRENCE WOLFF, M. D. 3d edition, enlarged by some 300 Essential Formulæ, selected from eminent authorities, by WM. M. POWELL, M. D. (Double number, price $2.00.)
10. **ESSENTIALS OF GYNÆCOLOGY.** 3d edition, revised. With 62 illustrations. By EDWIN B. CRAGIN, M. D.
11. **ESSENTIALS OF DISEASES OF THE SKIN.** 3d edition, revised and enlarged. 71 letter-press cuts and 15 half-tone illustrations. By HENRY W. STELWAGON, M. D. (Price, $1.00 net.)
12. **ESSENTIALS OF MINOR SURGERY, BANDAGING, AND VENEREAL DISEASES.** 2d edition, revised and enlarged. 78 illustrations. By EDWARD MARTIN, M. D.
13. **ESSENTIALS OF LEGAL MEDICINE, TOXICOLOGY, AND HYGIENE.** 130 illustrations. By C. E. ARMAND SEMPLE, M. D.
14. **ESSENTIALS OF DISEASES OF THE EYE, NOSE, AND THROAT.** 124 illustrations. 2d edition, revised. By EDWARD JACKSON, M. D., and E. BALDWIN GLEASON, M. D.
15. **ESSENTIALS OF DISEASES OF CHILDREN.** 2d edition. By WILLIAM M. POWELL, M. D.
16. **ESSENTIALS OF EXAMINATION OF URINE.** Colored "VOGEL SCALE," and numerous illustrations. By LAWRENCE WOLFF, M. D. (Price, 75 cents.)
17. **ESSENTIALS OF DIAGNOSIS.** By S. SOLIS-COHEN, M. D., and A. A. ESHNER, M. D. 55 illustrations, some in colors. (Price, $1.50 net.)
18. **ESSENTIALS OF PRACTICE OF PHARMACY.** By L. E. SAYRE. 2d edition, revised and enlarged.
20. **ESSENTIALS OF BACTERIOLOGY.** 3d edition. 82 illustrations. By M. V. BALL, M. D.
21. **ESSENTIALS OF NERVOUS DISEASES AND INSANITY.** 48 illustrations. 2d edition, revised. By JOHN C. SHAW, M. D.
22. **ESSENTIALS OF MEDICAL PHYSICS.** 155 illustrations. 2d edition, revised. By FRED J. BROCKWAY, M. D. (Price, $1.00 net.)
23. **ESSENTIALS OF MEDICAL ELECTRICITY.** 65 illustrations. By DAVID D. STEWART, M. D., and EDWARD S. LAWRANCE, M. D.
24. **ESSENTIALS OF DISEASES OF THE EAR.** By E. B. GLEASON, M. D. 89 illustrations.

JUST PUBLISHED.

A TEXT-BOOK OF MATERIA MEDICA, THERAPEUTICS, AND PHARMACOLOGY. By GEORGE F. BUTLER, PH. G., M. D., Professor of Materia Medica and of Clinical Medicine in the College of Physicians and Surgeons, Chicago; Professor of Materia Medica and Therapeutics, Northwestern University, Woman's Medical School, etc. 8vo, 858 pages. Illustrated. Prices: Cloth, $4.00 net; Sheep or Half-Morocco, $5.00 net

A clear, concise, and practical text-book, adapted for permanent reference no less than for the requirements of the class-room. The arrangement (embodying the synthetic classification of drugs based upon therapeutic affinities) is believed to be at once the most philosophical and rational, as well as that best calculated to engage the interest of those to whom the academic study of the subject is wont to offer no little perplexity.

Special attention has been given to the Pharmaceutical section, which is exceptionally lucid and complete.

LECTURES ON RENAL AND URINARY DISEASES. By ROBERT SAUNDBY, M. D. Edin., Fellow of the Royal College of Physicians, London, and of the Royal Medico-Chirurgical Society; Physician to the General Hospital; Consulting Physician to the Eye Hospital and to the Hospital for Diseases of Women; Professor of Medicine in Mason College, Birmingham, etc. 8vo, 434 pages, with numerous illustrations and 4 colored plates. Price, Cloth, $2.50 net.

In these Lectures, which are a re-issue in one volume of the author's well-known works on *Bright's Disease* and *Diabetes*, there is given, within a modest compass, a review of the present state of knowledge of these important affections, with such additions and suggestions as have resulted from the author's thirteen years' clinical and pathological study of the subjects. The lectures have been carefully revised and much new matter added to them. There has also been added a section dealing with "Miscellaneous Affections of the Kidney," making the book more complete as a work of reference.

ELEMENTARY BANDAGING AND SURGICAL DRESSING, with Directions concerning the Immediate Treatment of Cases of Emergency. For the use of Dressers and Nurses. By WALTER PYE, F. R. C. S., late Surgeon to St. Mary's Hospital, London. Small 12mo, with over 80 illustrations. Cloth, flexible covers. Price, 75 cents net.

This little book is chiefly a condensation of those portions of Pye's "Surgical Handicraft" which deal with bandaging, splinting, etc., and of those which treat of the management in the first instance of cases of emergency. Within its own limits, however, the book is complete, and it is hoped that it will prove extremely useful to students when they begin their work in the wards and casualty rooms, and useful also to surgical nurses and dressers.

"The directions are clear and the illustrations are good."—*London Lancet.*

"The author writes well, the diagrams are clear, and the book itself is small and portable, although the paper and type are good."—*British Medical Journal.*

"One of the most useful little works for dressers and nurses. The author truly says that it is 'a very little book,' but it is large in usefulness."—*Chemist and Druggist.*

JUST ISSUED. SOLD BY SUBSCRIPTION.

ANOMALIES
AND
CURIOSITIES OF MEDICINE.
BY
GEORGE M. GOULD, M. D.,
AND
WALTER L. PYLE, M. D.

Several years of exhaustive research have been spent by the authors in the great medical libraries of the United States and Europe in collecting the material for this work. **Medical literature of all ages and all languages** has been carefully searched, as a glance at the Bibliographic Index will show. The facts, which will be of **extreme value to the author and lecturer,** have been arranged and annotated, and full reference footnotes given, indicating whence they have been obtained.

In view of the persistent and dominant interest in the anomalous and curious, a **thorough and systematic collection** of this kind (the first of which the authors have knowledge) must have its own peculiar sphere of usefulness.

As a complete and authoritative **Book of Reference** it will be of value not only to members of the medical profession, but to all persons interested in general scientific, sociologic, and medico-legal topics; in fact, the general interest of the subject and the dearth of any complete work upon it make this volume **one of the most important literary innovations of the day.**

An especially valuable feature of the book consists of the **Indexing.** Besides a complete and comprehensive **General Index,** containing numerous cross-references to the subjects discussed, and the names of the authors of the more important reports, there is a convenient **Bibliographic Index** and a **Table of Contents.**

The plan has been adopted of printing the **topical headings in bold-face type,** the reader being thereby enabled to tell at a glance the subject-matter of any particular paragraph or page.

Illustrations have been freely employed throughout the work, there being 165 relief cuts and 130 half-tones in the text, and 12 colored and half-tone full-page plates—a total of over 320 separate figures.

The careful rendering of the text and references, the wealth of illustrations, the mechanical skill represented in the typography, the printing, and the binding, combine to make this book one of the most attractive medical publications ever issued.

Handsome Imperial Octavo Volume of 968 Pages.
PRICES: Cloth, $6.00 net; Half Morocco, $7.00 net.

JUST ISSUED.

PENROSE'S DISEASES OF WOMEN.
A Text-Book of Diseases of Women. By CHARLES B. PENROSE, M. D., PH.D., Professor of Gynecology, University of Pennsylvania; Surgeon to the Gynecean Hospital, Philadelphia. Octavo volume of 529 pages, handsomely illustrated. Price, $3.50 net.

MALLORY AND WRIGHT'S PATHOLOGICAL TECHNIQUE.
Pathological Technique. By FRANK B. MALLORY, A. M., M. D., Asst. Professor of Pathology, Harvard University Medical School; and JAMES H. WRIGHT, A. M., M. D., Instructor in Pathology, Harvard University Medical School. Octavo volume of 396 pages, handsomely illustrated.

SENN'S GENITO-URINARY TUBERCULOSIS.
Tuberculosis of the Genito-Urinary Organs, Male and Female. By NICHOLAS SENN, M. D., PH.D., LL.D., Professor of the Practice of Surgery and of Clinical Surgery, Rush Medical College, Chicago. Handsome octavo volume of 320 pages. Illustrated.

SUTTON AND GILES' DISEASES OF WOMEN.
Diseases of Women. By J. BLAND SUTTON, F. R. C. S., Asst. Surgeon to Middlesex Hospital, and Surgeon to Chelsea Hospital, London; and ARTHUR E. GILES, M. D., B. Sc. Lond., F. R. C. S. Edin., Asst. Surgeon to Chelsea Hospital, London. 436 pages, handsomely illustrated. Price, $2.50 net.

IN PREPARATION.

ANDERS' PRACTICE OF MEDICINE.
A Text-Book of the Practice of Medicine. By JAMES M. ANDERS, M. D., PH.D., LL.D., Professor of the Practice of Medicine and of Clinical Medicine, Medico-Chirurgical College, Philadelphia. *In press.*

AN AMERICAN TEXT-BOOK OF GENITO-URINARY AND SKIN DISEASES.
Edited by L. BOLTON BANGS, M. D., Late Professor of Genito-Urinary and Venereal Diseases, New York Post-Graduate Medical School and Hospital, and WILLIAM A. HARDAWAY, M. D., Professor of Diseases of the Skin, Missouri Medical College.

AN AMERICAN TEXT-BOOK OF DISEASES OF THE EYE, EAR, NOSE, AND THROAT.
Edited by G. E. DE SCHWEINITZ, M. D., Professor of Ophthalmology in the Jefferson Medical College, and B. ALEXANDER RANDALL, M. D., Professor of Diseases of the Ear in the University of Pennsylvania and in the Philadelphia Polyclinic.

MACDONALD'S SURGICAL DIAGNOSIS AND TREATMENT.
Surgical Diagnosis and Treatment. By J. W. MACDONALD, M. D., Graduate of Medicine of the University of Edinburgh; Licentiate of the Royal College of Surgeons, Edinburgh; Professor of the Practice of Surgery and of Clinical Surgery, Minneapolis College of Physicians and Surgeons.

HIRST'S OBSTETRICS.
A Text-Book of Obstetrics. By BARTON COOKE HIRST, M. D., Professor of Obstetrics, University of Pennsylvania.

MOORE'S ORTHOPEDIC SURGERY.
A Manual of Orthopedic Surgery. By JAMES E. MOORE, M. D., Professor of Orthopedics and Adjunct Professor of Clinical Surgery, University of Minnesota, College of Medicine and Surgery.

HEISLER'S EMBRYOLOGY.
A Text-Book of Embryology. By JOHN C. HEISLER, M. D., Prosector to the Professor of Anatomy, Medical Department of the University of Pennsylvania.

NOW READY—VOLUMES FOR 1896 AND 1897.

SAUNDERS'
American Year-Book of Medicine and Surgery
COLLECTED AND ARRANGED BY EMINENT AMERICAN SPECIALISTS AND TEACHERS,

UNDER THE EDITORIAL CHARGE OF

GEORGE M. GOULD, M. D.

The knowledge gained is equivalent to a post-graduate course.

NOTWITHSTANDING the rapid multiplication of medical and surgical works, still these publications fail to meet fully the requirements of the *general physician*, inasmuch as he feels the need of something more than mere text-books of well-known principles of medical science. Mr. Saunders has long been impressed with this fact, which is confirmed by the unanimity of expression from the profession at large, as indicated by advices from his large corps of canvassers.

This deficiency would best be met by current journalistic literature, but most practitioners have scant access to this almost unlimited source of information, and the busy practiser has but little time to search out in periodicals the many interesting cases, whose study would doubtless be of inestimable value in his practice. Therefore, a work which places before the physician in convenient form *an epitomization of this literature by persons competent to pronounce upon*

The Value of a Discovery or of a Method of Treatment

cannot but command his highest appreciation. It is this critical and judicial function that will be assumed by the Editorial staff of the "American Year-Book of Medicine and Surgery."

It is the special purpose of the Editor, whose experience peculiarly qualifies him for the preparation of this work, not only to review the contributions to American journals, but also the methods and discoveries reported in the leading medical journals of Europe, thus enlarging the survey and making the work characteristically **international.** These reviews will not simply be a series of undigested abstracts indiscriminately run together, nor will they be retrospective of "news" *one or two years old*, but the treatment presented will be *synthetic* and *dogmatic*, and will include **only what is new.** Moreover, through expert condensation by experienced writers, these discussions will be

Comprised in a Single Volume of about 1200 Pages.

The work will be replete with **original** and **selected** illustrations skilfully reproduced, for the most part, in Mr. Saunders' own studios established for the purpose, thus ensuring accuracy in delineation, affording efficient aids to a right comprehension of the text, and adding to the attractiveness of the volume.

Prices: Cloth, $6.50 net; Half Morocco, $7.50 net.

W. B. SAUNDERS, Publisher,
925 Walnut Street, Philadelphia.

www.ingramcontent.com/pod-product-compliance
Lightning Source LLC
Chambersburg PA
CBHW032352230426
43672CB00007B/675